# OPTICAL MATERIALS

# OPTICAL ENGINEERING

*Series Editor*

**Brian J. Thompson**
William F. May Professor of Engineering
*and* Provost, University of Rochester
Rochester, New York

*Laser Engineering Editor:* **Peter K. Cheo**
United Technologies Research Center
Hartford, Connecticut

*Other Volumes in Preparation*

# OPTICAL MATERIALS

## AN INTRODUCTION TO
## SELECTION AND APPLICATION

**Solomon Musikant**
Paoli, Pennsylvania

MARCEL DEKKER, INC.          New York and Basel

Library of Congress Cataloging in Publication Data

Musikant, Solomon.
  Optical materials.

  (Optical engineering ; 6)
  Includes bibliographies and index.
  1. Optical materials. I Title. II. Series: Optical
engineering (Marcel Dekker, Inc.) ; v. 6.
QC374.M87    1985    621.36    84-28704
ISBN 0-8247-7309-8

MARCEL DEKKER, INC.
270 Madison Avenue, New York, New York  10016

Current printing (last digit):
10  9  8  7  6  5  4  3  2  1

PRINTED IN THE UNITED STATES OF AMERICA

# ABOUT THE SERIES

Optical science, engineering, and technology have grown rapidly in the last decade so that today optical engineering has emerged as an important discipline in its own right. This series is devoted to discussing topics in optical engineering at a level that will be useful to those working in the field or attempting to design systems that are based on optical techniques or that have significant optical subsystems. The philosophy is not to provide detailed monographs on narrow subject areas but to deal with the material at a level that makes it immediately useful to the practicing scientist and engineer. These are not research monographs, although we expect that workers in optical research will find them extremely valuable.

Volumes in this series cover those topics that have been a part of the rapid expansion of optical engineering. The developments that have led to this expansion include the laser and its many commercial and industrial applications, the new optical materials, gradient index optics, electro- and acousto-optics, fiber optics and communications, optical computing and pattern recognition, optical data reading, recording and storage, biomedical instrumentation, industrial robotics, integrated optics, infrared and ultraviolet systems, etc. Since the optical industry is currently one of the major growth industries this list will surely become even more extensive.

Brian J. Thompson
University of Rochester
Rochester, New York

# PREFACE

The aim of this book is to provide the optics designer or user
with information on the broad range of materials used as optical
elements in systems and devices. For each class of materials
(glasses, crystalline materials, plastics, coatings) fundamental
performance requirements, basic characteristics, principles of
fabrication, possibilities for new or modified materials, and key
characterization data are provided.

This volume will give the reader a broad perspective on the
optical materials now available and the possibilities for their future
development, as well as provide data useful for preliminary materials selections and optical design. The bulk of the discussion
relates to refracting materials, because this is the area of great-
est variety and most generalized applicability in the optic arts.

The contents are intended to be useful to a wide range of
readers, including

Optical system and device designers and developers
Optics designers and optics engineers
Materials engineers
Physical measurements engineers
Test engineers
Students of optical sciences
Students of materials sciences

It is hoped that these practitioners will find the organization
of the information to be a helpful aid in appreciating and therefore properly evaluating the various materials available for specific applications.

The book is designed mainly for the technical worker who
already has become knowledgeable in one or more aspects of optical phenomena, applications, and/or materials science.

The volume, furthermore, is intended to be practical and not
mathematically involved.

Solomon Musikant

# ACKNOWLEDGMENTS

I wish to thank my many co-workers and collaborators at the General Electric Company, especially Drs. R. J. Charles, S. Prochazka, G. A. Slack and R. A. Tanzilli, and those at the Society of Photo-Optical Instrumentation Engineers, particularly R. E. Fischer, for the stretching of my knowledge engendered by these associations. In addition, I am grateful to the various university and U.S. government scientific personnel who have supported and encouraged the optical materials research work in which I have been engaged. Although this is indeed a long list, I wish to express special appreciation to Drs. A. L. Bement, H. E. Bennett, A. M. Diness, W. A. Harrison, W. T. Messick, R. Pohanka, E. vanReuth, R. Rice, and W. B. White.

Finally, in this review of acknowledgments, warm thanks go to my wife, Shirley, who encouraged me throughout the preparation of this book.

Solomon Musikant

# CONTENTS

Contents

# 1
# TRANSMISSION, REFLECTION, AND ABSORPTION OF LIGHT

## 1.1 ELECTROMAGNETIC SPECTRUM

The true nature of light is probably impossible to know. However, Maxwell's theory (James Clerk Maxwell, 1831–1879) and the quantum theory provide a consistent theoretical explanation of all optical phenomena. Maxwell assumed that light is merely one form of electromagnetic energy that has a wave form and a periodic nature.

These electromagnetic waves travel at a fixed velocity in a given medium and have a range of frequencies, or wavelengths. In a vacuum, electromagnetic waves propagate at a velocity $v_0$ of $3 \times 10^{10}$ cm/s. The relationship among wavelength $\lambda$, frequency $\nu$, and velocity in the medium, v, is given by

$$\lambda = \frac{v}{\nu}$$

where $\lambda$ is the wavelength (cm), $\nu$ the frequency (Hz), and v the velocity (cm/s).

The units of length commonly used when discussing wave motion in the optical region are

| Unit | Symbol | Length |
| --- | --- | --- |
| Micrometer | μm | $10^{-6}$ m |
| Nanometer | nm | $10^{-9}$ m |
| Angstrom | Å | $10^{-10}$ m |

The unit of frequency is the hertz (cycles per second) abbreviated Hz.

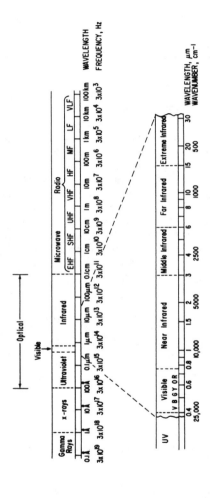

**Figure 1.1** Electromagnetic spectrum. (Reprinted with permission of F. Grum and R. J. Becherer, *Optical Radiation Measurements*, Academic Press, New York, 1979, Vol. 1.)

The electromagnetic spectrum is partitioned into various classes of electromagnetic waves, based on ranges of wavelength. Visible electromagnetic waves (light) extend from 400 to 750 nm. Electromagnetic waves with wavelengths down to 10 nm are called ultraviolet light. The lower limit is a matter of definition. The infrared band extends from 750 to $10^6$ nm ($10^3$ $\mu$m). The electromagnetic spectrum designations are shown in Fig. 1.1.

We will be directing our discussion to the portion of the spectrum bounded by the ultraviolet (UV) region on the short-wavelength side to the infrared (IR) on the long-wavelength side.

## 1.2 BAND STRUCTURE OF METALS, DIELECTRICS, AND SEMICONDUCTORS

Quantum theory was developed during the early part of this century by Planck, Einstein, Bohr, de Broglie, Schrödinger, and Heisenberg. This theory accurately predicts the behavior of a wide range of solid state phenomena. An excellent introductory treatment of this subject was given by Bohm (1951).

The quantum theory employs Newton's idea that light is composed of small discrete bodies, now called photons. Modern theory accepts the dual nature of light, i.e., that light exhibits both the wave character of Maxwell's formulations and the particle character envisaged by Newton. Maxwell's theory treats the propagation of light, while the quantum theory deals with the interaction of light with matter.

Photon energy can be described in terms of wavelength $\lambda$, wave number N, joules (J), electron volts (eV), or frequency. These are related through the following equations:

$$v = \lambda \nu \text{ cm/s}$$
$$E = h\nu \text{ J} \quad \text{or} \quad E = 0.6242 \times 10^{19} \, h\nu \text{ eV}$$
$$N = 1/\lambda \text{ cm}^{-1}$$

where v is the velocity in the medium (cm/s), n the frequency (Hz), E the energy of photon (J or eV), h = $6.62517 \times 10^{-34}$ J/s (Planck's constant), N is the wave number (cm$^{-1}$) or number of waves in 1 cm of path, and 1 eV = $1.602 \times 10^{-19}$ J. Frequency, wavelength, and quantum energy in various regions of the electromagnetic spectrum are shown in Table 1.1.

Consider a crystal to be made up of a collection of atoms in a regular array, known as a lattice. Each atom has one or more electrons in its outer electron shell. In an isolated atom, the electrons surrounding the nucleus are restricted to a set of discrete energy levels as required by the quantum theory.

**Table 1.1**  Frequency, Wavelength, and Quantum Energy for Various Bands in the Electromagnetic Spectrum[a]

| Type of radiation | | Frequency | Wavelength | Quantum Energy |
|---|---|---|---|---|
| Wave region | radio waves | $10^9$ Hz and less | 300 mm and longer | 0.000004 eV and less |
| | microwaves | $10^9$ to $10^{12}$ Hz | 300 to 0.3 mm | 0.000004 to 0.004 eV |
| Optical region | infrared | $10^{12}$ to $4.3 \times 10^{14}$ Hz | 300 to 0.7 μm | 0.004 to 1.7 eV |
| | visible | $4.3 \times 10^{14}$ to $5.7 \times 10^{14}$ Hz | 0.7 to 0.4 μm | 1.7 to 2.3 eV |
| | ultraviolet | $5.7 \times 10^{14}$ to $10^{16}$ Hz | 0.4 to 0.03 μm | 2.3 to 40 eV |
| Ray region | x rays | $10^{16}$ to $10^{19}$ Hz | 300 to 0.3 Å | 40 to 40,000 eV |
| | gamma rays | $10^{19}$ Hz and above | 0.3 Å and shorter | 40,000 eV and above |

[a]From G. R. Fowles, *Introduction to Modern Optics*, Holt, Rinehart and Winston, Inc., New York, 1968. (Reprinted by permission of Holt, Rinehart and Winston, CBS College Publishing.)

Each level can hold up to two electrons. The lower levels are occupied and the higher ones become occupied as the lower energy electrons become excited. In a crystal, the single discrete levels of the individual atoms broaden into a band of closely spaced discrete energy levels.

Figure 1.2 shows a number of band configurations. The ordinate represents energy level. The abscissa can be thought of as the distance from the surface of the crystal. The dark regions represent bands that are filled with electrons. The lighter regions represent unoccupied allowable (quantum) levels. The white regions represent forbidden energy values. That is, in accordance with the quantum theory, no electrons may occupy any energy level in the forbidden bands.

Electrical conduction can take place if the electrons can be excited into an unfilled quantum level. Thus in Fig. 1.2a, a very small increment of energy or excitation will promote the lower (valence) electrons into the empty (conduction) band. Thus Fig. 1.2a represents the band configuration for a metallic conductor.

Figure 1.2b represents the band configuration for an insulator or dielectric material. A tremendous amount of energy must be imparted to the electrons in the valence band to promote them to the conduction band. The difference in energy $E_g$ between the top of the valence band and the bottom of the conduction band is designated as the band gap. $E_g$ values of various materials are shown in Table 1.2.

Semiconductors are characterized by a small band gap as indicated in Fig. 1.2c. In this case, $E_g$ is low and small excitations promote the electrons into the conduction band, leaving unfilled positions in the valence band which then becomes conductive by movement of the "holes" left behind. In an intrinsic semiconductor, such as germanium, the number of conducting electrons equals the

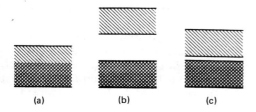

(a)          (b)          (c)

Figure 1.2   Electron occupation in various band configurations: (a) conductor, (b) insulator (dielectric), and (c) semiconductor (from Hutchinson and Baird, 1963).

Table 1.2    Values of the Energy Gap Between the Valence and
Conduction Bands at Room Temperature[a]

| Crystal | $E_g$ (eV) | Crystal | $E_g$ (eV) |
|---------|-----------|---------|-----------|
| Diamond | 5.33 | ZnS | 3.6 |
| Si | 1.14 | ZnSe | 2.60 |
| GaAs | 1.4 | AsCl | 3.2 |
| SiC | 3 | AsI | 2.8 |
| $Al_2O_3$ | 8+ | $TiO_2$ | 3 |
| Ge | 0.75 | | |

[a]Source:  Kittel (1966).

number of holes.  However, by a process known as "doping," an
extrinsic semiconductor can be fabricated to contain a preponder-
ance of electron type conductors upon excitation or a preponder-
ance of hole type conductors upon excitation.  The former is
called an n-type semiconductor and the latter, p-type.

A fuller treatment of this area can be found in many textbooks
on solid state theory such as Kittel (1966), Smith (1961), and
Hutchinson and Baird (1963).  An excellent text on materials science has been written by van Vlack (1964).

For metals, the band gap is low so that photons of either energy, UV to IR, will excite the metal electronic structure and
thus induce reemission of energy.

In the visible, this reemitted energy is observed by the eye
as metallic lustre.  Semiconductors have an intermediate level of
$E_g$ and thus will tend to be excited by photons of an intermediate
energy content and be transparent for lower energy photons,
i.e., exhibit lustre in the visible and be transparent in the IR,
where no interaction of the photon and the electronic structure
takes place.

If there is no interaction between the photons of a particular
energy level and the electronic structure of the material, then
the material is transparent to the photon.  Dielectrics have a high
$E_g$ and only high-energy photons in the UV can excite the electronic structure.  Therefore, dielectrics tend to be transparent
in the visible and IR regimes.

In addition to the electronic excitation, photons also excite
the ionic lattice.  This excitation reveals itself as lattice absorption of photons of relatively low energy in the IR and short millimeter radio frequency bands.

## 1.3 OPTICAL PROPERTIES OF MATERIALS

The intensity I of an electromagnetic wave is the amount of energy per unit time transmitting through a unit area which is transverse to the propagation direction.

A collection of atoms (a material) interacts with incoming photons (a light beam). The atoms and associated electronic structure which are excited by the photons reemit photons. The propagation pattern which results is observed in terms of surface reflection and transmission characteristics of the optical material.

The optical properties of the material are functions of the wavelength of the incoming light, temperature of and applied pressure on the material, as well as the environment (including electromagnetic fields) acting on the material.

The index of refraction n of a material which transmits light is given by

$$n = \frac{v_0}{v}$$

where $v_0$ is the velocity of light in a vacuum and v the velocity of light in the medium.

The complex refractive index n* includes the parameter $\kappa$, which is related to the energy absorbed by the medium from the light beam as it propagates through the medium. The complex refractive index is given by (Jenkins and White, 1957)

$$n^* = n(1 - i\kappa) \quad \text{and} \quad \kappa = \frac{\alpha\lambda}{4\pi n}$$

where $\kappa$ is the absorption index and $\alpha$ is the absorption coefficient.

Figure 1.3 characterizes metals, semiconductors, and dielectrics in terms of the index of absorption as a function of wavelength. The specific shape and location of absorption peaks varies with the nature of the specific material involved.

Consider Fig. 1.4, which represents a light beam propagating through a solid. The intensity $I_0$ of the beam just after entering the medium is attenuated as it passes through the solid medium. That attenuation is given by Bouguer's law (Pierre Bouguer, 1698–1758), which states

$$I = I_0 e^{-\alpha x}$$

where I is the intensity at position x, x the distance along the path, $\alpha$ the absorption coefficient, and e the base of natural logarithms.

The optical region of Fig. 1.3 is shown in an expanded plot for a typical dielectric in Fig. 1.5. In this case, the data are plotted

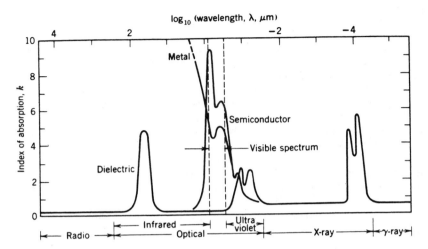

**Figure 1.3**  Frequency variation of the index of absorption κ for metals, semiconductors, and dielectrics. (Reprinted with permission from W. D. Kingery, H. K. Bowen, and D. R. Uhlmann, *Introduction to Ceramics*, John Wiley and Sons, Inc., New York, 1976.)

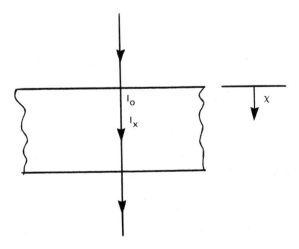

**Figure 1.4**  Transmission of a light beam in a solid.

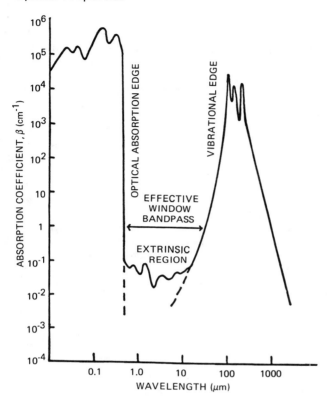

**Figure 1.5**   Optical bandpass for a dielectric (from Musikant et al., 1980).

as absorption coefficient versus wavelength. The optical absorption edge on the short-wavelength side of the diagram is a consequence of the band structure of the material and is an intrinsic characteristic of the structure of the material, as is the vibrational edge on the long-wavelength side. Here the absorption is due to the interaction of the low-energy photons with the lattice structure elasticity. However, in the effective bandpass region there generally exist extrinsic absorptions due to imperfections. These imperfections may be impurity atoms, unintentional second phases, voids, nonhomogeneous composition, and, in multiphase and polycrystalline materials, boundary effects.

Another important distinction when interpreting transmission and absorption data is that between specular and nonspecular transmission. In specular transmission, the light is either specularly ($\theta = \theta'$) reflected from the boundaries of the optical element, absorbed, or transmitted in a straight line ($\theta = \theta''$). However, in nonspecular transmission, the light is scattered as it passes through the optical element and thus the emerging beam is diffuse, as illustrated in Fig. 1.6.

When making measurements in a spectrophotometer, only light entering a solid angle $\phi$, characteristic of the instrument, is observed by the detector. For precise imaging, only a narrow spreading of the beam can be tolerated. This narrow beam is termed a specular beam. However, when accounting for the energy being emitted from the optical element, all the energy in the diffuse envelope has to be considered.

One of the most important goals in development of an optical refractive material is the minimization of all features which lead to absorption and scattering of the incoming electromagnetic signal. A continuing challenge to the materials scientist is the elimination of the extrinsic absorbers and scatterers to reduce unwanted

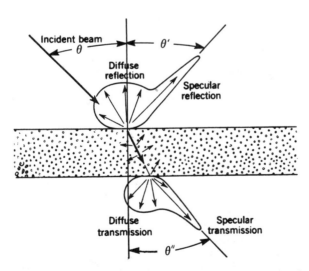

Figure 1.6  Nonspecular transmission. (Reprinted with permission from W. D. Kingery, H. K. Bowen, and D. R. Uhlmann, *Introduction to Ceramics*, John Wiley and Sons, Inc., New York, 1976.)

attenuations to the minimum level possible, i.e., the intrinsic
state.

Another significant point is that similar intrinsic and extrinsic
effects occur in materials useful both in the optical and radio-
frequency (rf) regimes.  For example, Fig. 1.7 shows an intrinsic
absorption curve for a typical dielectric material.  In this case, the
ordinate is absorption coefficient and the abscissa is wave number.
If a maximum absorption coefficient of $10^{-1}$ cm$^{-1}$ is specified, then
from considerations of intrinsic properties the material, even in its
theoretically purest form, is inappropriate for application at wave
numbers $N_1 < N < N_2$.  Absorption coefficient data in the optical

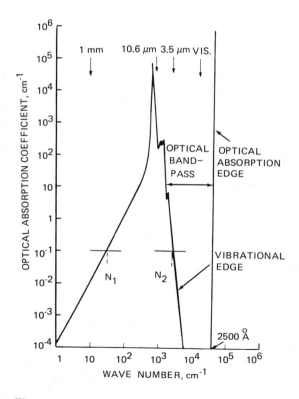

Figure 1.7   Absorptance of a typical dielectric material (AℓN)
(from Musikant et al., 1980).  Original calculation by G. A. Slack,
General Electric Co.

regime are known, whereas the data for the rf regime are usually not known for most materials.

## 1.4 TUNING OF LATTICE VIBRATIONS

A covalent crystal is one in which the atoms are strongly connected by two electrons, one from each participating atom. Covalent bonds are highly directional. The ionic elastic vibrations of a crystalline refractive material can be modeled in terms of the atomic mass and the force constant required to displace each atom from its equilibrium position in the lattice. The fundamental natural frequency of the lattice for a covalent crystal is given by

$$\nu^2 = 2k(1/M_c + 1/M_a)$$

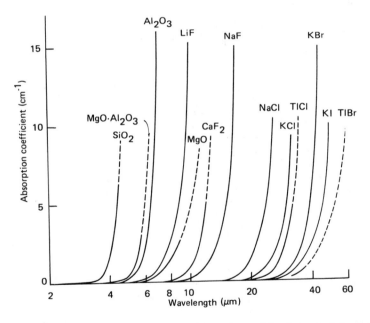

Figure 1.8   Infrared absorption edge of ionic crystals. (Reprinted with permission from W. D. Kingery, H. K. Bowen, and D. R. Uhlmann, *Introduction to Ceramics*, John Wiley and Sons, Inc., New York, 1976.)

where $\nu$ is the natural frequency, k the elastic constant for small displacement of the ion, $M_c$ the ionic mass of the cation, and $M_a$ the ionic mass of the anion. The natural frequency of the crystal is given by

$$\nu = \sqrt{2k/\mu}$$

where $\mu$ is the reduced mass of the cation–anion pair given by

$$1/\mu = (1/M_c + 1/M_a)$$

The phonon absorption band starts at the lattice fundamental, where the photon energy is absorbed and converted to elastic vibration. For a long-wave cutoff a low value of k and a low value of the term $1/\mu$ are desired. Thus the halide salts have a long cutoff as indicated in Fig. 1.8, which is a display of the absorption coefficients versus wavelength for a variety of ionic crystals. The long cutoff

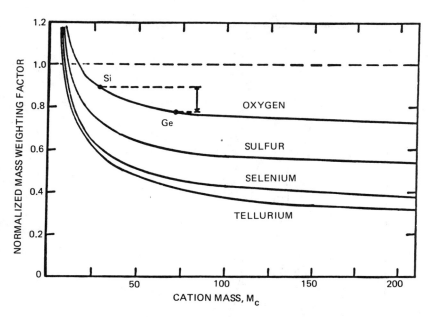

Figure 1.9   Mass weighting factors for various anions versus cation mass $M_c$. [S. Musikant, Development of a new family of improved infrared (IR) dome ceramics, in *Emerging Optical Materials*, Proc. SPIE 297, 1981, pp. 2–12.] Original calculation by W. White, Pennsylvania State University.

versus wavelength for a variety of ionic crystals. The long cutoff
of the halide salts is a direct consequence of the low elastic constants. Unfortunately, this low elastic constant is also associated
with weak and soft substances.

Figure 1.9 shows a normalized mass weighting factor as a function of $M_c$ for various binary ionic compounds involving oxygen,
sulfur, selenium, and tellurium as the anions. The implication of
the chart is that as one moves downward and to the right in Fig.
1.6, the reduced mass $\mu$ increases and $\nu$ decreases, but the effects
are much stronger for increases of cation mass at the low atomic
number portion of the Periodic Table for a given anion and for increases in anion mass.

Figure 1.10 illustrates the increased cut-off $\lambda$ for $GeO_2$ glass
(cation atomic weight 72.59) compared to $SiO_2$ glass (cation atomic
weight 28.09).

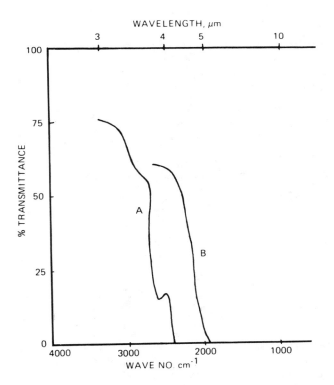

Figure 1.10   Transmission data for (A) $SiO_2$ glass and (B) $GeO_2$
glass.

## 1.5 BASIC OPTICAL RELATIONS

An awareness of certain elementary optical relationships is of
critical importance when selecting or developing optical materials.
An excellent introductory treatise on optics has been given by
Meyer-Arendt (1972).  A few key principles will be reviewed in
the following.

The index of refraction of a substance was defined in Sec. 1.3
as the inverse ratio of the velocity of light in that substance relative to the velocity of light *in vacuo* at a specific wavelength and
at a specific temperature and pressure.

For solids the index is generally measured at 1 atm or lower
pressures.  Strain birefringence in solids is discussed in Section
2.4.  The refractive index n is a function of wavelength $\lambda$, with
n generally decreasing as $\lambda$ increases from the UV to the IR
regime.

The dispersion at any $\lambda$ is defined by

$$(\text{dispersion})_\lambda = \left(\frac{dn}{d\lambda}\right)_\lambda$$

For optical glasses, it is customary to quote values of the refractive index at a number of characteristic spectral lines, most
commonly:

| | | |
|---|---|---|
| Hydrogen F | 4861 Å |
| Helium d | 5876 Å |
| Sodium D | 5893 Å |
| Hydrogen C | 6563 Å |

Typical dispersion curves for three generic glasses are shown
in Fig. 1.11.  Similar data for ceramic crystalline materials are
shown in Fig. 1.12.  The Abbe number is an index of the degree
of dispersion and is defined in terms of the above-mentioned spectral lines as

$$\nu_d = \frac{n_d - 1}{n_F - n_C}$$

The Abbe number is a measure of the chromatic aberration in an
optical material.  In general a high value of $\nu_d$ and a high value of
n are desirable if a single optical element is to serve as an effective
refracting device with minimal chromatic aberration.  However, in
practice, a large variety of glasses have been developed for the
myriad designs of optical engineers and scientists.  Figure 1.13

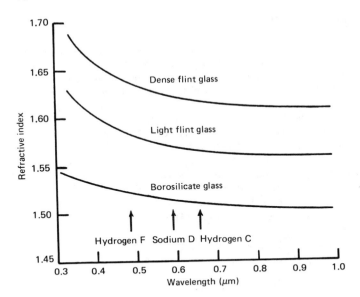

**Figure 1.11** Refractive index versus wavelength for glasses (showing dispersion). (Reprinted with permission from W. D. Kingery, H. K. Bowen, and D. R. Uhlmann, *Introduction to Ceramics*, John Wiley and Sons, Inc., New York, 1976.)

displays a typical map of $\nu_d$ versus $n_d$ of more or less traditional glasses which are available commercially (a) and some crystalline ceramics (b).

A refractive element transmits electromagnetic radiation and in general the maximization of the transmission is desired by the designer. Boughuer's law, stated earlier, defines the absorption coefficient $\alpha$ of a material.

The absorption coefficient $\alpha$ is a physical property of the medium and is generally a function of $\lambda$ and temperature. The absorption coefficient is also a sensitive function of the exact nature of the medium, i.e., it is affected by compositional variations, impurities, and inhomogeneities.

Partial reflection of the incident beam occurs at the boundary of two materials with different values of n. Referring to Fig. 1.14, for near normal incidence of unpolarized light on a refractive element in vacuum, the reflectance r, or fraction of incident light which is reflected, for a single surface is given by

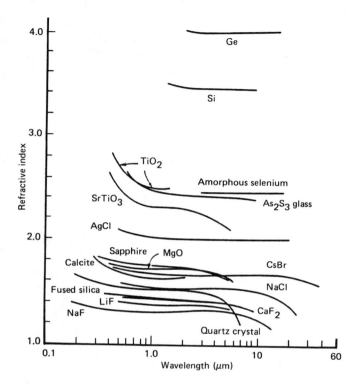

**Figure 1.12**   Refractive index versus wavelength for crystalline solids (showing dispersion). (Reprinted with permission of W. L. Wolfe and G. J. Zissis, *The Infrared Handbook*, Environmental Research Institute of Michigan, Ann Arbor, 1978.)

$$r = [(n - 1)/(n + 1)]^2$$

Generally, r is a function of the angle of incidence. For incident angles greater than 30° the incident angle and the polarization of the light complicate the physical result and complex analytical expressions are required. The theoretical treatments were originally set forth by Augustin Jean Fresnel (1788–1827) and such reflections are called Fresnel reflections.

The ratio of the primary transmitted wave intensity to the incident wave intensity is known as the external transmittance. The ratio of these intensities just before leaving the medium and just after entering the medium is the internal transmittance.

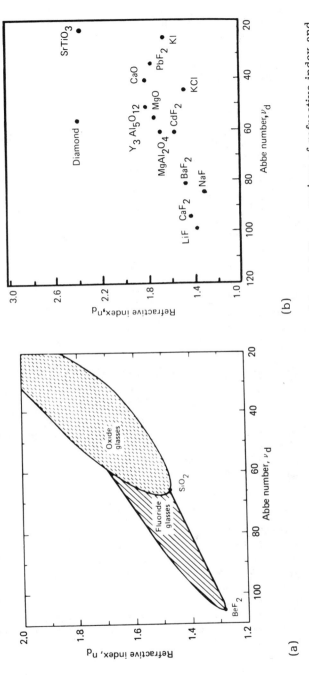

**Figure 1.13** $n_d$ versus $\nu d$ Abbe number for optical materials. (a) Known regions of refractive index and reciprocal dispersion (Abbe number) of optical glasses. (b) Refractive index and Abbe number for cubic crystals. (From M. J. Weber, D. Milam, and W. L. Smith, Nonlinear refractive index of glasses and crystals. Opt. Eng. 17(5), 1978.)

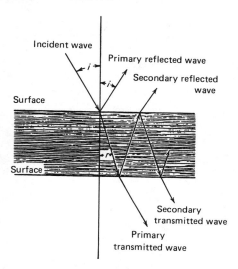

**Figure 1.14** Reflectance of a light beam. (Reprinted with permission from W. D. Kingery, H. K. Bowen, and D. R. Uhlmann, *Introduction to Ceramics*, John Wiley and Sons, Inc., New York, 1976.)

As mentioned earlier, in addition to the bulk absorption coefficient $\alpha$, there is an additional attenuation by scattering of the specular beam due to inhomogeneities and Fresnel reflections at the grain boundaries of polycrystalline noncubic solids. In noncubic birefringent solids, the index of refraction varies depending on the orientation of the crystallographic axes to the light beam. This birefringent effect gives rise to scattering.

The reduction in the intensity of the specular or nonscattered beam due to scattering can be included in an expanded Bouguer's law as

$$I_0/I = e^{-(\alpha_i + \alpha_s)x}$$

where $\alpha_i$ is the intrinsic absorption coefficient and $\alpha_s$ the scattering absorption coefficient.

For materials which exhibit significant scattering, this point must be considered carefully when designing and predicting the performance of optical elements. Typical values of average refractive index and some birefringence data in the visible are presented in Table 1.3.

There are many excellent introductory texts on optical theory, including Jenkins and White (1957) and Fincham and Freeman (1980).

**Table 1.3**  Refractive Index and Birefringence Data for Solids[a]

|  | Average refractive index | Birefringence |
|---|---|---|
| **Glass composition** | | |
| From orthoclase ($KAlSi_3O_8$) | 1.51 | |
| From albite ($NaAlSi_3O_8$) | 1.49 | |
| From nepheline syenite | 1.50 | |
| Silica glass, $SiO_2$ | 1.458 | |
| Vycor glass (96% $SiO_2$) | 1.458 | |
| Soda-lime−silica glass | 1.51−1.52 | |
| Borosilicate (Pyrex) glass | 1.47 | |
| Dense flint optical glasses | 1.6−1.7 | |
| Arsenic trisulfide glass, $As_2S_3$ | 2.66 | |
| **Crystals** | | |
| Silicon chloride, $SiCl_4$ | 1.412 | |
| Lithium fluoride, LiF | 1.392 | |
| Sodium fluoride, NaF | 1.326 | |
| Calcium fluoride, $CaF_2$ | 1.434 | |
| Corundum, $Al_2O_3$ | 1.76 | 0.008 |
| Periclase, MgO | 1.74 | |
| Quartz, $SiO_2$ | 1.55 | 0.009 |
| Spinel, $MgAl_2O_4$ | 1.72 | |
| Zircon, $ZiSiO_4$ | 1.95 | 0.055 |
| Orthoclase, $KAlSi_3O_8$ | 1.525 | 0.007 |
| Albite, $NaAlSi_3O_8$ | 1.529 | 0.008 |
| Anorthite, $CaAl_2Si_2O_8$ | 1.585 | 0.008 |
| Sillimanite, $Al_2O_3 \cdot SiO_2$ | 1.65 | 0.021 |
| Mullite, $3Al_2O_3 \cdot 2SiO_2$ | 1.64 | 0.010 |

Table 1.3 (Continued)

| | Average refractive index | Birefringence |
|---|---|---|
| Rutile, $TiO_2$ | 2.71 | 0.287 |
| Silicon carbide, SiC | 2.68 | 0.043 |
| Litharge, PbO | 2.61 | |
| Galena, PbS | 3.912 | |
| Calcite, $CaCO_3$ | 1.65 | 0.17 |
| Silicon, Si | 3.49 | |
| Cadmium telluride, CdTe | 2.74 | |
| Cadmium sulfide, CdS | 2.50 | |
| Strontium titanate, $SrTiO_3$ | 2.49 | |
| Lithium niobate, $LiNbO_3$ | 2.31 | |
| Yttrium oxide, $Y_2O_3$ | 1.92 | |
| Zinc selenide, ZnSe | 2.62 | |
| Barium titanate, $BaTiO_3$ | 2.40 | |

[a]Reprinted with permission from W. D. Kingery, H. K. Bowen, and D. R. Uhlmann, *Introduction to Ceramics*, John Wiley and Sons, Inc., New York, 1976.

## REFERENCES

D. Bohm, *Quantum Theory*, Prentice-Hall, Inc., Englewood Cliffs, NJ, 1951.

W. H. A. Fincham and M. H. Freeman, *Optics*, Butterworth Publishers, Inc., Woburn, MA, 1980.

G. R. Fowles, *Introduction to Modern Optics*, Holt, Rinehart and Winston, Inc., New York, 1968.

F. Grum and R. J. Becherer, *Optical Radiation Measurements*, Academic Press, New York, 1979.

T. S. Hutchinson and D. C. Baird, *The Physics of Engineering Solids*, John Wiley and Sons, Inc., New York, 1963.

F. A. Jenkins and H. E. White, *Fundamentals of Optics*, McGraw-Hill Book Publishing Co., Inc., New York, 1957.

W. D. Kingery, H. K. Bowen, and D. R. Uhlmann, *Introduction to Ceramics*, John Wiley and Sons, Inc., New York, 1976.

C. Kittel, *Introduction to Solid State Physics*, John Wiley and Sons, Inc., New York, 1966.

J. R. Meyer-Arendt, *Introduction to Classical and Modern Optics*, Prentice-Hall, Inc., Englewood Cliffs, NJ, 1972.

S. Musikant, *Development of a New Family of Improved Infrared (IR) Dome Ceramics*, Proceedings Vol. 297, SPIE, Bellingham, WA, 1981.

S. Musikant, R. A. Tanzilli, H. Rauch, S. Prochazka, and I. C. Huseby, Advanced optical ceramics, In *Proceedings of the 15th Symposium on Electromagnetic Windows*, June 18–20, 1980, Georgia Institute of Technology, Atlanta, GA, 1980.

R. A. Smith, *Semiconductors*, Cambridge University Press, Cambridge, England, 1961.

L. H. van Vlack, *Elements of Materials Science*, Addison-Wesley Publishing Co., Inc., Reading, MA, 1964.

M. J. Weber, D. Milam, W. L. Smith, Nonlinear refractive index of glasses and crystals, Opt. Eng. *17*(5), 1978.

W. L. Wolfe and G. J. Zissis, Eds., *The Infrared Handbook*, Environmental Research Institute of Michigan, Ann Arbor, MI, 1978.

# 2
# GLASS

## 2.1 INTRODUCTION

Both modern glass and modern optics grow largely from one of the great historic coincidences, the coming together in the town of Jena in Germany in the year 1880 of three great giants—the university physicist Ernst Abbe, the lensmaker Carl Zeiss, and the ceramic scientist Otto Schott.

Abbe wrote a book *Refracting and Dispersing Power of Solid and Fluid Bodies* and formulated the physical requirements for optical glasses which were needed to design a variety of high quality and novel photographic and scientific instruments. Schott performed the synthesis and ceramic research and development to determine which glass compositions and treatments could provide these properties. Zeiss, the master craftsman, fashioned these new glasses into lenses and optical elements and created the new tools from which our current technological explosion was launched.

## 2.2 GLASS—GENERAL CHARACTERISTICS

Glass may be defined as an amorphous undercooled inorganic liquid. At room temperature the viscosity is so high that the material is rigid in a practical sense.

Devitrification is defined as the formation of a crystalline phase when glass is held at a critical temperature for a long enough time. The main problems of the glassmaker are to achieve a homogeneous amorphous body by a suitable procedure and the avoidance of devitrification during any of the thermal processes which the material must undergo in transforming the raw ingredients into an optically useful solid.

For example, in any given system of oxides, only a restricted range of composition will permit the existence of a glassy phase.

Figure 2.1 illustrates the compositional region within which glasses can be formed for the following systems:

$SiO_2 - Al_2O_3 - Li_2O$

$SiO_2 - Al_2O_3 - ZnO$

$SiO_2 - Al_2O_3 - MgO$

$SiO_2 - ZnO - Li_2O$

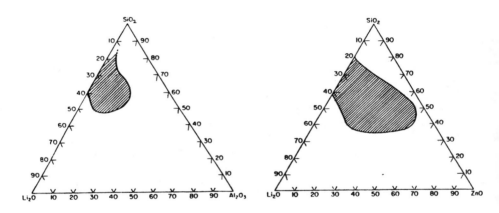

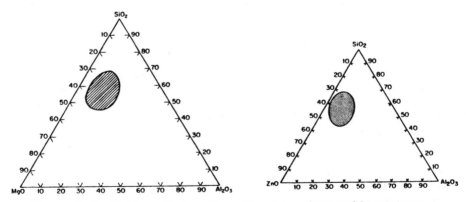

Figure 2.1   Region of glass formation in various oxide systems. (Reprinted with permission of P. W. McMillan, *Glass Ceramics*, Academic Press, New York, 1964.)

Outside the compositions represented by the shaded areas no glass will form without the possibility of substantial or total devitrification during the cooling process.

The resultant glass must have the optical and physical properties and chemical durability suitable for the intended application.

A list of critical items regarding procurement of optical glass is shown in Fig. 2.2. The tabulation is not all-inclusive, since many

1.  NOMINAL OPTICAL PROPERTIES

    Refractive index

    Abbe number

    Partial dispersion

    Internal transmittance

2.  CHEMICAL PROPERTIES

    Resistance to climate, staining acid and alkalinity

3.  THERMAL PROPERTIES

    Viscosity

    Thermal expansion

    Index variation with respect to temperature

4.  MECHANICAL PROPERTIES

    Hardness

    Strength

    Modulus of elasticity

5.  QUALITY

    Deviation of index of refraction and Abbe No.

    Homogeneity

    Striae

    Bubbles

    Stress birefringence

Figure 2.2   Critical items regarding the procurement of optical glass.

specialized requirements arise from a great variety of optical devices which need glass elements.

Optical inhomogeneity can be created by the development of internal stress in the body. Annealing is a thermal process in which the body is heated to a temperature high enough to allow stress equilibration to be achieved followed by a controlled cooling cycle which limits internal stress to an acceptably low level.

Aside from the optical properties desired, glass must be fluid enough at an industrially accessible temperature to be melted on a commercial scale and devitrification resistant so that it can be worked above its freezing point at relatively high viscosity yet be of such a composition that the body does not devitrify during the subsequent thermal annealing steps.

## 2.3 GLASS COMPOSITIONS

The properties of glass are determined primarily by its formulation and secondarily by the details of the processing to which that formulation is subjected.

The common oxides used in formulating glasses and some of their functions are listed in Table 2.1.

In addition to these basic ingredients of glass, many other components are used to impart special properties. For example,

Table 2.1 Common Oxides Used in Formulating Glasses

| Oxide | Function | Cost |
|---|---|---|
| $SiO_2$ | Principal glass former | Cheapest material |
| $Al_2O_3$ | Adds durability | Fairly expensive |
| CaO | Enhances glass formation | Inexpensive |
| MgO | Usually reduces melting temperature | Fairly expensive |
| $B_2O_3$ | Reduces viscosity, reduces melting temperature, does not affect durability | Very expensive |
| $Na_2O$ | Reduces viscosity, decreases durability | Second most expensive |

PbO is commonly used in optical glass to increase the index of refraction.

Silica is the outstanding glass forming oxide because of its resistance to devitrification, resistance to attack by water and most acids, and low coefficient of expansion. The most effective flux (i.e., substance which depresses melting point) for $SiO_2$ is $Na_2O$, usually added as sodium carbonate $Na_2CO_3$. Lime ($CaO$) is added as an inexpensive ingredient which imparts chemical durability and enhances glass formation. However, an excess of lime will make the glass hard to melt and lead to devitrification.

Alumina ($Al_2O_3$) improves chemical durability, decreases the coefficient of expansion, reduces the tendency toward devitrification but increases viscosity. Schott systematically explored the effects of $Al_2O_3$ on optical properties.

## Optical Glass Compositions

A typical composition for window glass of the early twentieth century is

| | |
|---|---|
| $SiO_2$ | 71.5 wt% |
| $Al_2O_3$ | 1.5 |
| $Na_2O$ | 14.0 |
| CaO | 13.0 |

Although window glass is an optical material, the demands on its optical properties are minimal. For glasses to be useful in optical systems, a relatively wide range of materials must be available. Abbe number and index of refraction are the key parameters and, since the early researches of Schott, many optical glasses have been developed. These glasses have traditional designators as shown in Table 2.2. The symbol $R$ represents an alkali metal ion or an alkaline earth metal ion in the formulas.

The flints have a high index of refraction and a low Abbe number, while the crowns have a low index of refraction and a high Abbe number. The compositions indicated in Table 2.2 are generic in nature. The PbO increases the index of refraction. Boron oxide, alumina, rare earths, and, recently, fluorides all have been employed to achieve a variety of combinations of index of refraction and Abbe number.

Tables 2.3 and 2.4 illustrate a variety of typical optical glasses which have been developed for commercial applications. The possible glass compositions are virtually infinite.

**Table 2.2**  Traditional Designations for Optical Glasses

| | Type and components of optical glasses | |
|---|---|---|
| SF | dense flint | |
| F | flint | |
| LF | light flint | $SiO_2$ PbO $R_2O^a$ |
| LLF | extra light flint | |
| KF | crown flint | |
| K | crown | $SiO_2$ RO $R_2O$ |
| BaSF | dense barium flint | $SiO_2$ PbO BaO $R_2O$ |
| BaF | barium flint | $SiO_2$ $B_2O_3$ PbO BaO |
| BaLF | light barium flint | |
| BaK | barium crown | $SiO_2$ BaO $R_2O$ |
| BaLK | light barium crown | |
| SSK | extra dense barium crown | $SiO_2$ $B_2O_3$ BaO |
| SK | dense barium crown | |
| LaF | lanthanum flint | $(SiO_2)$ $B_2O_3$ $La_2O$ PbO $Al_2O_3$ |
| LaLF | light lanthanum flint | $(SiO_2)$ $B_2O_3$ $La_2O_3$ ObO RO |
| LaLK | light lanthanum crown | $(SiO_2)$ $B_2O_3$ $La_2O_3$ $ZrO_2$ RO |
| LaK | lanthanum crown | |
| TaSF | dense tantalum flint | $B_2O_3$ $La_2O_3$ $ThO_2$ $Ta_2O_5$ |
| TaF | tantalum flint | |
| Tak | tantalum crown | $B_2O_3$ $La_2O_3$ $ThO_2$ RO |
| BK | borosilicate crown | $SiO_2$ $B_2O_3$ $R_2O$ BaO |
| PK | phosphate crown | |
| PSK | dense phosphate crown | $Al_2O_3$ $P_2O_5$ RO |
| FK | fluor crown | $SiO_2$ $BO_3$ $R_2O$ RF |

[a] R denotes alkaline earth (Group IIa) or alkali metal (Group Ia).

A current chart (Fig. 2.3) by Ohara Optical Glass Co. illustrates the varieties of glass available from a single supplier.

Table 2.3 Some Optical Glasses and Their Properties[a,b]

| | Refractive index at D line $n_D$ | Mean dispersion[d] $n_F - n_C$ | Reciprocal relative dispersion $\nu$ | Partial dispersions and ratios[c] | | |
|---|---|---|---|---|---|---|
| | | | | $n_D - n_C$ | $n_F - n_D$ | $n_{G'} - n_F$ |
| 1. Borosilicate crown | 1.51700 | 0.00801 | 64.5 | 0.00239 (0.298) | 0.00199 (0.248) | 0.00447 (0.558) |
| 2. Crown | 1.52300 | 0.00893 | 58.6 | 0.00264 (0.296) | 0.00220 (0.246) | 0.00506 (0.567) |
| 3. Light barium crown | 1.54110 | 0.00904 | 59.9 | 0.00268 (0.296) | 0.00222 (0.246) | 0.00511 (0.568) |
| 4. Dense barium crown | 1.61700 | 0.01123 | 54.9 | 0.00330 (0.294) | 0.00277 (0.247) | 0.00649 (0.578) |
| 5. Extra dense barium crown | 1.65611 | 0.01294 | 50.7 | 0.00378 (0.292) | 0.00322 (0.249) | 0.00751 (0.580) |
| 6. Crown flint | 1.52860 | 0.01024 | 51.6 | 0.00300 (0.193) | 0.00252 (0.246) | 0.00594 (0.580) |
| 7. Light barium flint | 1.56210 | 0.01103 | 51.0 | 0.00321 (0.292) | 0.00272 (0.247) | 0.00632 (0.574) |
| 8. Barium flint | 1.60530 | 0.01388 | 43.6 | 0.00400 (0.238) | 0.00342 (0.246) | 0.00827 (0.596) |

Table 2.3  (Continued)

| | Refractive index at D line $n_D$ | Mean dispersion[d] $n_F - n_C$ | Reciprocal relative dispersion $\nu$ | Partial dispersions and ratios[c] | | |
|---|---|---|---|---|---|---|
| | | | | $n_D - n_C$ | $n_F - n_D$ | $n_{G'} - n_F$ |
| 9. Dense barium flint | 1.61700 | 0.01601 | 38.6 | 0.00468 (0.286) | 0.00393 (0.245) | 0.00968 (0.605) |
| 10. Extra light flint | 1.55850 | 0.01227 | 45.5 | 0.00355 (0.289) | 0.00302 (0.246) | 0.00725 (0.591) |
| 11. Light flint | 1.57250 | 0.01347 | 42.5 | 0.00389 (0.289) | 0.00331 (0.246) | 0.00803 (0.596) |
| 12. Dense flint | 1.61700 | 0.01686 | 36.6 | 0.00482 (0.286) | 0.00415 (0.246) | 0.01025 (0.608) |
| 13. Extra dense flint | 1.75060 | 0.02707 | 27.7 | 0.00758 (0.290) | 0.00663 (0.245) | 0.01707 (0.631) |
| 14. Phosphate flint | 1.98020 | 0.00417 | 22.2 | 0.01215 (0.275) | 0.01075 (0.243) | 0.02876 (0.651) |

[a]Source:  Tooley (1960).
[b]Courtesy, Bausch & Lomb Incorporated.
[c]Ratios in parentheses are $(n_D - n_C)/(n_F - n_C)$, $(n_F - n_D)/(n_F - n_C)$, and $(n_{G'} - n_F)/(n_F - n_C)$.
[d]Mean dispersion calculated from $(1 - n_D)/(n_F - n_C)$.

**Table 2.4** Compositions of Glasses in Table 2.3a,b

| No. | $SiO_2$ | $B_2O_3$ | $Al_2O_3$ | $K_2O$ + $Na_2O$ | $CaO$[c] | $BaO$[d] | $PbO$ | $ZnO$ | $ZrO_2$ + $TiO_2$ | $Sb_2O_3$ + $As_2O_3$ |
|---|---|---|---|---|---|---|---|---|---|---|
| 1 | 72.5 | 12.5 | – | 12 | – | 1 | – | 2 | – | (0.2) |
| 2 | 73 | 1 | – | 15 | 10 | – | – | 1 | – | (0.2) |
| 3[e] | 71 | 5 | – | 11 | – | 8.5 | – | 3 | – | (0.2) |
| 4 | 57.5 | 6 | 4.5 | – | – | 25 | 0.5 | 6 | 0.5 | (0.3) |
| 5[e] | 50.5 | 12.5 | 1.5 | – | – | 24 | – | 3.5 | 6 | – |
| 6 | 81 | – | – | 11 | – | – | 3 | 3.5 | – | 1.5 |
| 7 | 70 | 2.5 | – | 10 | – | 7 | 4 | 6.5 | – | (0.3) |
| 8 | 66 | – | – | 7 | – | 9 | 9 | 9 | – | (0.2) |
| 9 | 69.5 | – | – | 9 | – | 4.5 | 16 | 1 | – | (0.1) |
| 10 | 77 | 1 | – | 11 | – | 1 | 10 | – | – | (0.1) |
| 11 | 76 | – | – | 11 | – | 1 | 12 | – | – | (0.1) |
| 12 | 73.5 | – | – | 6.5 | – | – | 19.5 | 0.5 | – | (0.1) |
| 13 | 62.5 | – | – | 2 | – | – | 34.5 | – | – | (0.5) |
| 14[e] | – | – | – | – | – | – | 43 | – | 2 | – |

[a] Source: Tooley (1960).
[b] Compositions are in mole percent (M%).
[c] Some MgO included.
[d] Some SrO included.
[e] No. 3 contains 1.5% $Li_2O$; No. 5, 2% $La_2O_3$; No. 14, 15% $WO_3$, 38% $P_2O_5$, 2% CdO.

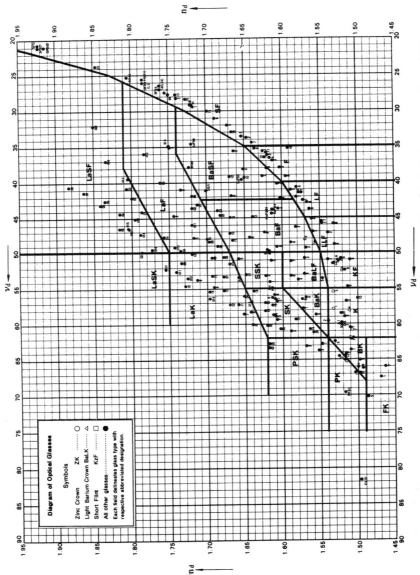

Figure 2.3  $n_d$ versus Abbe number for optical glasses. [Reprinted by permission of Ohara Optical Glass, Inc.; from Ohara Optical Glass, Inc. (1982).]

## 2.4 STRAIN BIREFRINGENCE

The fact that strain induces birefringence in glass was discovered
by Sir David Brewster in 1813. Birefringence is expressed as the
difference between the path length traversed by rays of light propagating in the direction of maximum strain and those propagating
in the transverse direction per unit path length. This value is
conventionally expressed in nm/cm path length to characterize a
given state of strain.

Strain birefringence arises because of changes in interatomic
dimensions and distortion of the outer electrons around the ions.
For high index glasses (flints) the former effect predominates, while
for low index glasses (crowns), the latter. The stress optical coefficient is defined as

$$c = \frac{\text{change in optical path length}}{\text{unit stress}}$$

Typically for glasses we have $c = 3 \times 10^{-7}$ cm/kg cm$^{-2}$. Thus for
an optical path-length difference (OPD) of 30 nm due to strain birefringence, the stress can be deduced from

$$c = \frac{30 \times 10^{-9}}{\text{stress}}$$

$$\text{stress} = \frac{30 \times 10^{-9}}{3 \times 10^{-7}} = 10 \times 10^{-2} \text{ kg/cm}^2 \text{ (or } 10^{-1} \text{ kg/cm}^2)$$

Of course, the strain birefringence is undesirable for optical elements and therefore in the thermal processing of the optical glass
the development of strain is minimized by careful annealing. To
minimize strain exact control of the glass temperature is important
only during a short interval, the annealing range. At temperatures
greater than the annealing range the glass viscosity is so low that
no strain can persist. The lower limit of the annealing range, the
strain point, is that temperature from which the glass can be
quickly cooled without introducing permanent strain. Fine (or precise) annealing is used to reduce strain in optical glass.

## 2.5 VISCOSITY

When a unit cube of a fluid material is placed in a state of shear by
application of a shear force, as indicated in Fig. 2.4, the upper
face moves relative to the lower face. The viscosity is defined as

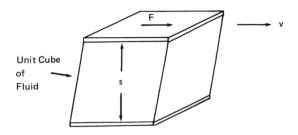

**Figure 2.4**    Unit cube of fluid in shear.

$$\eta = \frac{Fs}{v} \frac{(dyn/cm^2)(cm)}{cm/s}$$

where F is the shear force $(dyn/cm^2)$, s the height of the cube (cm), and v the velocity of top face relative to bottom face (cm/s). The unit of viscosity is known as the poise and is in units of $dyn/cm^2 \ s^{-1}$.

In working glass the viscosity is a key indicator to the behavior. Obviously, the viscosity increases as the temperature of the glass goes down from the melting point. Certain useful definitions relevant to viscosity are shown in Table 2.5. Annealing point and strain point ranges for typical glasses are given in Table 2.6.

**Table 2.5** Viscosity of Glass

|  | Viscosity (poise) |
|---|---|
| Flow point | $10^5$ |
| Softening point | $10^{7.6}$ |
| Upper limit of annealing range<br>Incipient softening point<br>Upper annealing temperature | $10^{11}$ to $10^{12}$ |
| Annealing point<br> temperature at which glass anneals<br> in 15 min | $10^{13.4}$ |
| Strain point<br> glass anneals in $\sim$ 16 hr<br> below this temperature there is<br> practically no viscous yield | $10^{14}$ |

Table 2.6   Annealing Point and Strain Point for Typical Glasses

|  | Annealing point (°C) | Strain point (°C) |
|---|---|---|
| Borosilicate glass | 518–550 | 470–503 |
| Lime glass | 472–523 | 412–474 |
| Lead glass | 419–451 | 353–380 |

## 2.6 ANNEALING

Annealing is a two-step process. The glass is raised to a temperature sufficiently high to allow removal of internal strain. It is then slowly cooled from the annealing temperature at a rate such that the permanent strain is low enough to be acceptable. A typical allowable birefringence is 5 nm/cm path. A typical annealing curve is shown in Fig. 2.5a,b. In Fig. 2.5b the $\log_{10}$ strain birefringence is plotted against time at a constant temperature. Curve fits show that an empirical relation of the form given below generally is appropriate to describe the behavior:

$$t = \frac{\ell}{A\,(\Delta n)}$$

where t is the annealing time (min), $\ell$ the thickness of slab (cm), n the strain birefringence (nm), and A the annealing constant (empirically determined from curve fit).

## 2.7 COMMERCIAL OPTICAL GLASS DESIGNATIONS

As mentioned earlier, each glass type is categorized by a group type such as BK (borosilicate crown) or SF (dense flint) followed by a six-digit number. The first three digits represent the refractive index $n_d$ and the second three, the Abbe number $\nu_d$. Thus glass BK511605 represents a glass with $n_d$ = 1.511 and $\nu_d$ = 0.00605. The refractive index is generally given in the medium range of the visible spectrum, usually $n_d$; the refractive index at λ = 587.56 nm, a spectral line of helium. The Abbe number is computed from

$$\nu_d = \frac{n_d - 1}{n_F - n_C}$$

where $n_d$ is the refractive index at 587.56 (He), $n_F$ the refractive index at 486.13 (H), and $n_C$ the refractive index at 656.28 (H).

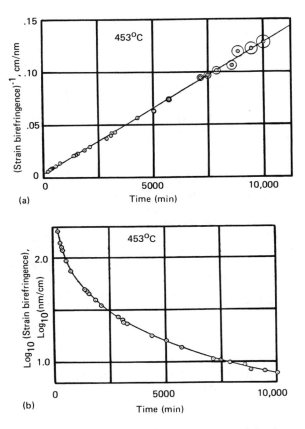

Figure 2.5  Typical annealing curve for glass. (a) Annealing curve, reciprocal of strain birefringence against time, of a glass at 453°C. (b) Annealing curve, logarithm of strain birefringence against time, using the same data as (a). (From Morey, 1954.)

The difference $n_F - n_C$ is called the "principal dispersion." Corresponding values at other wavelengths are also used, for example:

$$\nu_e = \frac{n_e - 1}{n_{F'} - n_{C'}}$$

where $n_e$ is the refractive index at 546.04 nm (Hg), $n_{F'}$ the refractive index at 479.99 nm (Cd), and $n_{C'}$ the refractive index at
643.85 (Cd). Refractive indices are generally measured for many
wavelengths as indicated in Table 2.7.

"Normal" glasses, as shown by Abbe, lie along a straight line
between types 511605 and 620363 on the $n_d$ versus $v_d$ plot. The
ordinate distance on the plot between other glasses and this normal line is called $\Delta P$. This quantity $\Delta P$ describes the dispersion
behavior of the "non-normal" glasses. Glasses which deviate from
the normal line are needed for correction of the secondary spectrum.

**Table 2.7**  Spectral Lines for Index of Refraction Determination[a]

| Wavelength (nm) | Designation | Spectral line | Element |
|---|---|---|---|
| 2325.4 | | Infrared mercury line | Hg |
| 1970.1 | | Infrared mercury line | Hg |
| 1529.6 | | Infrared mercury line | Hg |
| 1060.0 | | Neodymium glass laser | Nd |
| 1013.98 | t | Infrared mercury line | Hg |
| 852.1101 | s | Infrared cesium line | Cs |
| 706.5188 | r | Red helium line | He |
| 656.2725 | C | Red hydrogen line | H |
| 643.8469 | C' | Red cadmium line | Cd |
| 632.8 | | Helium—neon gas laser | He—Ne |
| 589.2938 | D | Yellow sodium line (center of the double line) | Na |
| 587.5618 | d | Yellow helium line | He |
| 546.0704 | e | Green mercury line | Hg |
| 486.1327 | F | Blue hydrogen line | H |
| 479.9914 | F' | Blue cadmium line | Cd |
| 435.8343 | g | Blue mercury line | Hg |
| 404.6561 | h | Violet mercury line | Hg |
| 365.0146 | i | Ultraviolet mercury line | Hg |

[a]Reprinted by permission of Schott Glass Technologies, Inc.;
source: Schott Glass Technologies, Inc. (1982).

The relative partial dispersion $P_{x,y}$ for wavelengths x and y can be defined by the following equation:

$$P_{x,y} = \frac{n_x - n_y}{n_F - n_C}$$

The so-called normal glasses obey the following linear relationship:

$$P_{x,y} \cong a_{xy} + b_{xy} \nu_d$$

where $a_{xy}$ and $b_{xy}$ are empirical constants derived from the normal glasses, which obey the line specified above.

An achromatic lens composed of two different glasses can be designed to focus any two colors in the same place, but all other colors will focus in slightly different positions. This small residual color defect is known as the secondary spectrum.

Correction of the secondary spectrum, i.e., achromatization for more than two wavelengths requires at least one glass that does not follow the normal glass linear rule. For a non-normal glass the relative partial dispersion is given by

$$P_{x,y} = a_{xy} + b_{xy} \nu_d + \Delta P_{xy}$$

where $a_{xy}$ and $b_{xy}$ are empirically determined for the normal glass for the relative partial dispersion of interest.

**Table 2.8**  Relative Partial Dispersions[a]

$$P_{C,t} = \frac{n_C - n_t}{n_F - n_C} \qquad P_{C,s} = \frac{n_C - n_s}{n_F - n_C} \qquad P_{F,e} = \frac{n_F - n_e}{n_F - n_C}$$

$$P_{g,F} = \frac{n_g - n_F}{n_F - n_C} \qquad P_{i,g} = \frac{n_i - n_g}{n_F - n_C}$$

Glasses K7 and F2 were selected as representative of the "normal glasses." The "normal lines" resulting from the pairs of values for K7 and F2 are shown below.

$\tilde{P}_{C,t} = 0.5450 + 0.004743 \, V_d$      $\tilde{P}_{g,F} = 0.6438 - 0.001682 \, V_d$

$\tilde{P}_{C,s} = 0.4029 + 0.002331 \, V_d$      $\tilde{P}_{i,g} = 1.7241 - 0.008382 \, V_d$

$\tilde{P}_{F,e} = 0.4884 - 0.000526 \, V_d$

[a]Reprinted by permission of Schott Glass Technologies, Inc.; source:  Schott Glass Technologies, Inc. (1982).

The $\Delta P_{x,y}$ values are tabulated for each type of glass for various relative partial dispersions. For example, Schott provides these deviations for the five relative partial dispersions shown in Table 2.8.

## 2.8 TEMPERATURE COEFFICIENT OF REFRACTION AND DISPERSION

The thermal coefficient of refraction dn/dT depends on the wavelength of light and the temperature. The absolute value is obtained by making measurements in vacuum, while the relative coefficient $(dn/dT)_{rel}$ is obtained by making the measurement at ambient air (760 Torr, dry air). The relationship between the relative and absolute thermal coefficients of refraction is expressed by

$$\left(\frac{dn}{dT}\right)_{abs} = \left(\frac{dn}{dT}\right)_{rel} + n \left(\frac{dn}{dT}\right)_{air}$$

where n is the refractive index of the glass at 1013.3 mbar (760 Torr).

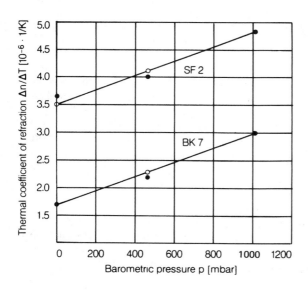

**Figure 2.6** Thermal coefficient of refraction as a function of barometric pressure (glasses SF2 and BK7); temperature range 20−40°C, wavelength $\lambda = 546.1$ nm. [Reprinted by permission of Schott Glass Technologies, Inc.; source: Schott Glass Technologies, Inc. (1982).]

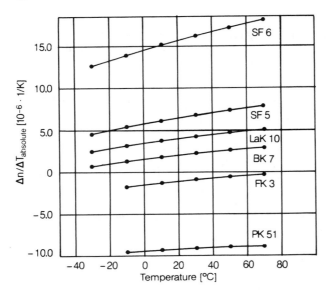

**Figure 2.7**   Absolute thermal coefficient of refraction of a number of optical glasses as a function of temperature; $\lambda$ = 435.8 nm. [Reprinted by permission of Schott Glass Technologies, Inc.; source:   Schott Glass Technologies, Inc. (1982).]

Figure 2.6 shows the thermal coefficient of refraction versus atmospheric pressure over the temperature range of +20 to +40°C and a wavelength of 546.1 nm for Schott glasses BK7 and SF2. Figures 2.7 and 2.8 show, respectively, the variation for BK7 and other glasses of the thermal coefficient of refraction as functions of temperature and wavelength.

## 2.9  CHEMICAL PROPERTIES

The resistance of glass to the various environments it may be exposed to varies depending on the glass composition.  Water, water vapor, acids, gases such as $CO_2$ and $SO_2$, as well as various ions in the polishing slurries, may all cause staining of the glass surface.  Staining is the optical effect produced by surface chemical reactions.

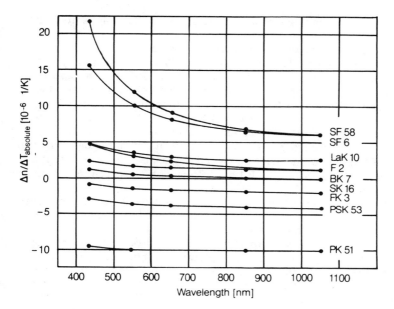

**Figure 2.8** Absolute thermal coefficient of refraction of a number of optical glasses as a function of wavelengths (temperature range, 20—40°C). [Reprinted by permission of Schott Glass Technologies, Inc.; source: Schott Glass Technologies, Inc. (1982).]

When glasses contain larger amounts of relatively insoluble components such as $SiO_2$, $Al_2O_3$, $TiO_2$ or oxides of the rare earths, they will tend to resist leaching by aqueous or acidic solutions. However, if a glass contains larger amounts of readily soluble compounds such as the alkali or alkaline earth oxides and the readily soluble oxides of boron and phosphorus, then one can expect reactions ranging from formation of coatings to dissolution of the glass surface.

## Climatic Resistance

The climatic resistance of optical glasses can be categorized by observing surface effects on light transmission after exposure to humidity testing. When water vapor is present in the air, particularly when the relative humidity and temperature are high, some glass surfaces can be attacked resulting in a cloudy surface film that cannot be removed easily. This cloudy film leads to light scattering.

In a typical test, polished glass is exposed to 100% relative humidity air and thermally cycled between 46 and 55°C hourly. After having been exposed for various times up to 180 hr, the specimens are measured for light scattering. Standard optical glasses (viz., Schott SK15, BaF, and F1) undergo surface changes sufficient to lead to scattering on the order of 2% to 5% of the incoming beam. Schott categorizes climatic resistance into four classes, CR1 to CR4, where CR1 represents glasses with few signs of deterioration after 180 hr of exposure and CR4 represents glasses prone to substantial effect, while CR2 and CR3 are representative ratings for intermediate glasses.

Under normal humidity conditions encountered during processing and storage, class CR1 glasses exhibit slight to no signs of deterioration. The other glasses require protection.

Ohara uses a somewhat different test. The test is carried out by putting freshly polished glass plates in a chamber with +50°C, 85% humidity for 24 hr. If the glass surface is severely attacked, another 6-hr test is carried out with new pieces. The classification in four groups, 1 through 4, is then obtained by inspecting the treated surface through a microscope with a magnification of 50, as shown in Table 2.9.

**Table 2.9**  Weathering Resistance Classification[a]

| Group | Classification |
|-------|----------------|
| 1 | When there is no fade on the glass exposed in the chamber for 24 hr and observed at 6000 luxes |
| 2 | When there is no fade observed on the glass exposed in the chamber for 24 hr at 1500 luxes but there is a fade observed at 6000 luxes |
| 3 | When a fade is observed on the glass exposed in a chamber for 24 hr when inspected at 1500 luxes |
| 4 | When a fade is observed on the glass exposed in a chamber for 6 hr when inspected at 1500 luxes |

[a]Reprinted by permission of Ohara Optical Glass, Inc.; source: Ohara Optical Glass, Inc. (1982).

## Staining Resistance

Resistance to staining relates to surface chemical effects due to
small quantities of slightly acidic water such as perspiration and
acidic condensation.  To test for this effect, two test solutions are
used, standard acetate (pH = 4.6) and sodium acetate buffer (pH =
5.6).  The test glass is exposed to test solutions, and the time
required to develop interference colors (staining) is observed.
Schott classifies stain resistance into classes FR0 to FR5, as indicated in Table 2.10.

## Acid Resistance

To assess acid resistance, more aggressive test solutions are used.
The time required to dissolve a 0.1 μm layer serves as a measure
of the resistance to acid attack.   Tables 2.11 and 2.12 indicate
the acid resistance classes SR1 to SR4 and SR51 to SR53 employed
by Schott.   Other optical glass manufacturers use equivalent
designations.

Ohara determines acid resistance of a polished surface by immers-
ing the polished glass element into N/2-nitric acid or acetic acid of
pH 4.6 at 25°C and measuring the time until the specific interfero-
metric color (purple) appears on the polished surface.  Acid resistance designations by Ohara are shown in Table 2.13.

Table 2.10  Staining Resistance Classes[a]

| Stain resistance classes[b] FR | 0 | 1 | 2 | 3 | 4 | 5 |
|---|---|---|---|---|---|---|
| Test solution | I[c] | I | I | I | II[d] | II |
| Time (hr) | 100 | 100 | 6 | 1 | 1 | 0.2 |
| Color change | no | yes | yes | yes | yes | yes |

[a]Reprinted by permission of Schott Glass Technologies, Inc.;
source:  Schott Glass Technologies, Inc. (1982).
[b]Allocation of optical glasses to stain resistance classes FRO—5
on the basis of the time that elapses before test solutions I and
II cause brown—blue staining (layer thickness ≈ 0.1 μm).
[c]Test solution I:  standard acetate, pH = 4.6.
[d]Test solution II:  sodium acetate buffer, pH = 5.6.

Table 2.11   Acid Resistance Classes[a]

| Acid resistance class[b] SR | 1 | 2 | 3 | 4 | 51—53 |
|---|---|---|---|---|---|
| Time (hr) | >100 | 100—10 | 10—1 | 1—0.1 | <0.1 |

[a]Reprinted by permission of Schott Glass Technologies, Inc.;
source:  Schott Glass Technologies, Inc. (1982).
[b]Classification of optical glasses by acid resistance classes SR1—
4 on the basis of the time required to dissolve a 0.1 µm layer
with an acidic solution of pH value 0.3 at 25°C.

Table 2.12   Acid Resistance Classes[a]

| Acid resistance class[b]SR | 51 | 52 | 53 |
|---|---|---|---|
| Time (hr) | 10—1 | 1—0.1 | <0.1 |

[a]Reprinted by permission of Schott Glass Technologies, Inc.;
source:  Schott Glass Technologies, Inc. (1982).
[b]Classification of sensitive optical glasses by acid resistance
classes SR51—53 on the basis of the time required to dissolve a
layer thickness of 0.1 µm in a weakly acidic solution of pH value
4.6 at 25°C.

Table 2.13   Acid Resistance Ratings[a]

| Group of acid resistance | In N/2-nitric acid | | | | In acetic acid of pH 4.6 | | |
|---|---|---|---|---|---|---|---|
| | 1 | 2 | 3 | 4 | 5a | 5b | 5c |
| Time (hr) | >100 | 100—11 | 10—2 | 1—0.1 | >1 | 1—0.1 | <0.1 |

[a]Reprinted by permission of Ohara Optical Glass, Inc.; source:
Ohara Optical Glass, Inc. (1982).

**Table 2.14** Alkaline Resistance Classes[a]

| Alkaline resistance[b] AR | 1 | 2 | 3 | 4 |
|---|---|---|---|---|
| Time (min) | >120 | 120−30 | 30−7.5 | <7.5 |

[a]Reprinted by permission of Schott Glass Technologies, Inc.; source: Schott Glass Technologies, Inc. (1982).
[b]Classification of optical glasses by alkaline resistance classes AR1−4 on the basis of the time required to decompose a layer of 0.1 μm thickness in a sodium hydroxide solution of pH value 10 at 90°C. The following key is used to characterize visible surface changes: .0, no change; .1, scarred surface, but no visible coatings (color change); .2, interference colors; .3, whitish staining; .4, white coating (thick layers).

### Resistance to Alkaline Solutions

Finishing processes such as grinding and polishing, which use water-based media, subject the glass to alkaline solutions because of chemical interaction of the water and the very fine abraded glass particles. To classify glasses as to alkaline resistance, the surface of the sample is exposed to sodium hydroxide of pH 10 at 90°C and the time in minutes to decompose an 0.1 μm layer is observed. Based on this type of observation, Schott classifies the glass into types AR1 to AR4 with the digit to the right of the decimal point representing the nature of the surface changes as noted in Table 2.14.

## 2.10 THERMAL EXPANSION OF GLASS

The dimensions and volume of glass increase as temperature is increased. A typical thermal expansion curve is shown in Fig. 2.9. Usually there is an initial nonlinear range A succeeded by a linear range B to the temperature at which the glass begins to become plastic and followed by the transformation region C. Beyond this the expansion coefficient $\alpha$ becomes much higher and achieves a second linear range. The transformation temperature $T_g$ is defined as indicated in Fig. 2.9 at the junction of the two linear expansion ranges.

Conventionally the thermal coefficient of expansion is reported for two temperature ranges, −30 to +70°C and +20 to 300°C. The $\alpha_{20-300°C}$ is in the range of $4 \times 10^{-6}$ to $16 \times 10^{-6}/°C$ for optical glasses.

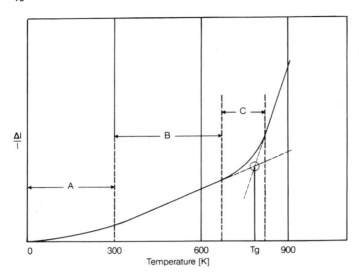

**Figure 2.9** Thermal expansion of glass. [Reprinted by permission of Schott Glass Technologies, Inc.; source: Schott Glass Technologies, Inc. (1982).]

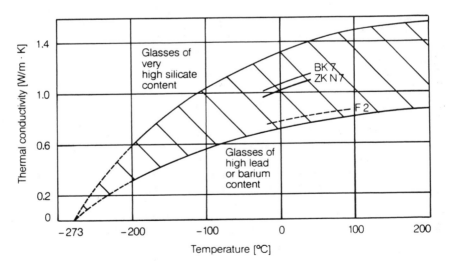

**Figure 2.10** Thermal conductivity of some optical glasses. [Reprinted by permission of Ohara Optical Glass, Inc.; source: Ohara Optical Glass, Inc. (1982).]

## 2.11   THERMAL CONDUCTIVITY

Optical glasses have a wide range of thermal conductivity; values range from 1.38 W/m K for fused silica to about 0.5 W/m K for high lead content glasses. Figure 2.10 shows typical ranges of thermal conductivity up to 200°C for various classes of glass and for a few specific Schott glasses.

## 2.12   DEFECTS IN GLASS

Defects in bulk glass are characterized by striae, optical inhomogeneity, stress birefringence, and bubbles. These are measured quantitatively and qualitatively, and the glasses are partitioned into classes with designations which define the quality of the glass.

Striae are localized stringy regions in glass with sufficient difference in index of refraction to be detectable by shadowgraph and/or the Foucault knife edge test. Striae grades are categorized on the basis of predetermined patterns. Normal quality optical glass can have fine striae, but "precision quality" glass contains no striae which can be detected by these means.

Optical homogeneity is measured by the degree to which refractive index varies within a glass block or melt. For good practice the maximum variation of the refractive index within a melt is $\pm 1 \times 10^{-4}$. However, modern optical glass manufacturers do considerably better. For example, Table 2.15A shows the various classes of homogeneity which are available from Schott. Wavefront deviations at $\lambda = 632.8$ nm and various thicknesses are given in Table 2.15B.

### Residual Strain Birefringence

Residual strains in glass depend on annealing conditions (cooling rate, temperature distribution), type of glass, and the dimensions. Stress birefringence is measured as the difference in optical path length and is stated in nanometers per centimeter (nm/cm).

High quality optical glass has very low values of residual stress as indicated in Table 2.16 for Schott glasses.

### Bubbles

Bubbles are eliminated during glass "fining," where the molten glass is held just above its melting point for a designated period of time in an atmosphere which minimizes gas absorption into the melt. The index of bubble content is the ratio of the observed cross sectional area of all bubbles in square millimeters observed

**Table 2.15** Homogeneity Classes for Optical Glasses[a]

A. Schott homogeneity groups for optical glasses

| Homogeneity group | Max. variation of $n_d$ value | Availability |
|---|---|---|
| H1 | $\pm 2 \times 10^{-5}$ | Within a selected melt |
| H2 | $\pm 5 \times 10^{-6}$ | Within a cut blank |
| H3 | $\pm 2 \times 10^{-6}$ | Within a cut blank, depending on dimensions |
| H4 | $\pm 1 \times 10^{-6}$ | Within a cut blank, but dependent on the type of glass and dimensions |

B. Wavefront deviations at $\lambda = 632.8$ nm and various thicknesses

| Homogeneity group | 10 mm glass thickness | 25 mm glass thickness | 50 mm glass thickness | 100 mm glass thickness |
|---|---|---|---|---|
| H1 | $0.32\lambda$ | $0.79\lambda$ | $1.58\lambda$ | $3.16\lambda$ |
| H2 | $0.16\lambda$ | $0.40\lambda$ | $0.79\lambda$ | $1.58\lambda$ |
| H3 | $0.06\lambda$ | $0.16\lambda$ | $0.32\lambda$ | $0.63\lambda$ |
| H4 | $0.03\lambda$ | $0.08\lambda$ | $0.16\lambda$ | $0.32\lambda$ |

[a]Reprinted by permission of Schott Glass Technologies, Inc.; source: Schott Glass Technologies, Inc. (1982)

Table 2.16 Limit Values for Stress Birefringence in Optical Glasses[a],[b]

| Glass types and dimensions | Stress birefringence | | |
|---|---|---|---|
| | Normal quality after fine annealing (nm/cm) | Normal quality after special annealing (NSK) or precision quality (nm/cm) | Normal quality after special annealing (NSK) or precision quality after special annealing (PSK) (nm/cm) |
| All types of glass up to dimensions of about 160 × 160 × 100 mm$^3$ | $\leq 10$ | $\leq 6$ | $\leq 4$ |
| All types of glass up to diam. ∿300 mm d∿60 mm[c] | $\leq 10$ | $\leq 6$ | $\leq 3$–4 depending on type of glass |
| All types of glass up to diam. ∿600 mm d∿80 mm | $\leq 10$–12 depending type of glass | $\leq 6$ | $\leq 4$ |
| Types of glass such as BK7, BaK4, SK16, LF5, F2, SF2 up to diam. ∿800 mm d∿100 mm | $\leq 12$ | $\leq 8$ | $\leq 5$ |
| Types of glass such as BK7, K5, F2, SF2 up to diam. ∿1000 mm d∿200 mm | $\leq 20$ | $\leq 12$ | $\leq 8$ |

[a]Limit values for stress birefringence in cut blanks of various dimensions and with various annealing processes.
[b]Reprinted by permission of Schott Glass Technologies, Inc.; source: Schott Glass Technologies, Inc. (1982).
[c]d represents thickness.

Table 2.17  Bubble Classes for Optical Glasses[a]

| Bubble class | Total area of all bubbles/inclusions $\geq 0.05$ mm per 100 $cm^3$ of glass, in $mm^2$ |
|---|---|
| B0 | 0—0.029 |
| B1 | 0.03—0.10 |
| B2 | 0.11—0.25 |
| B3 | 0.26—0.50 |

[a]Reprinted by permission of Schott Glass Technologies, Inc.; source:  Schott Glass Technologies, Inc. (1982).

in a volume of 100 $cm^3$ of glass.  All bubbles (and other inclusions) measuring >0.05 mm are included.  Bubble classes are defined by Schott as shown in Table 2.17.

## 2.13  MECHANICAL PROPERTIES

The key mechanical properties are the modulus of elasticity, torsional rigidity, Poisson's ratio, modulus of rupture, and Knoop or Vickers hardness.  Typical values for these properties for optical glasses are presented in Table 2.18.

## 2.14  INFRARED TRANSMITTING GLASSES

The infrared regime of interest extends from the edge of the visible ($\sim 0.7$ µm) to about 20 µm for a wide variety of applications including spectroscopy and sensors of various types.
    The long-wavelength cutoff for silica is at approximately 4.0 µm, depending on the definition of cutoff wavelength.  For our purpose, we define cutoff as occurring at 50% of the maximum transmittance for a given material.  Figure 2.11 shows the published transmittance for three grades of fused quartz and type BK7 Schott optical glass. For the fused quartz materials the cutoff ranges from $\sim 2.7$ to 3.7 µm.  As mentioned in Sec. 1.3, the infrared cutoff is determined by the vibrational frequency of the anion—cation bond.  By replacing silicon with the higher atomic weight germanium as the cation in a glass, the resonant frequency of the anion—cation can be extended to about 6 µm.

Table 2.18 Typical Mechanical Properties of Optical Glasses

| Glass type | Density (g/cm$^3$) | Knoop hardness (kg/mm$^2$) | Young's modulus (GPa) | Coeff. of thermal expansion (10$^{-6}$ C$^{-1}$) | Poisson's ratio | Tensile strength (MPa) |
|---|---|---|---|---|---|---|
| Silicate glass Corning code 9753 | 2.8 | 657 (100 g load) | 98.6 | 5.95 (25–300 C) | 0.28 | 48 (MOR) |
| Germanate glass Corning code 9754 | 3.6 | 560 (100 g load) | 84.1 | 6.2 | 0.29 | 49.9 (MOR) abraded sample |
| Calcium aluminate Barr & Stroud BS398 glass | 2.9 | 758 | 107 | 8.75 (20–300 C) | | 90 |
| Fused silica glass | 2.2 | 500 (200 g load) | 73.7 | 0.57 (0–200 C) | 0.17 | 49.6 (MOR) abraded sample |
| 100% germania glass | 3.65 | | 43.3 | | | |
| Borosilicate crown | 2.51 | 520 | 81.3 | 8.3 (20–300 C) | 0.21 | |
| Pyrex glass Corning 7740 | 2.23 | 418 (100 g load) | 62.7 | 3.25 | 0.20 | 42.1 |
| Fluoride glass (Hf, Ba, Th, F) | | 250 | 16.5 | 4.3 (50 C) 13.8 (250 C) | | 62 |
| Low expansion glass Ohara E6 | | 520 | 57.5 | 2.5 (100–300 C) | 0.195 | 38.9 |

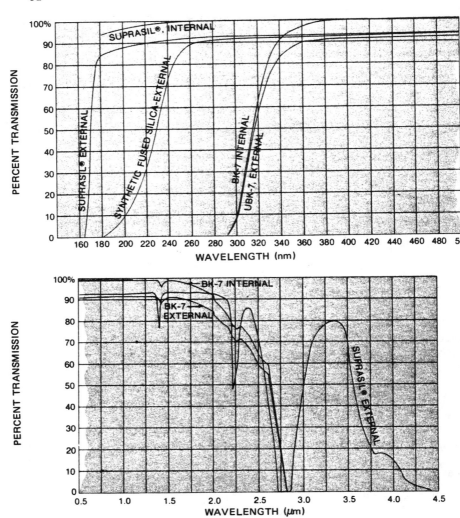

Figure 2.11   Transmittance of fused quartz.   Comparison at normal incidence of nominal uncoated external transmittances of various
optical materials for 10 mm sample thickness.   Limited surface loss
and internal transmittance data are also shown.   External transmittance here means useful external transmittance for image forming
purposes, i.e., single-pass reflection losses at both surfaces are
accounted for, while the effects of multiple reflection between the

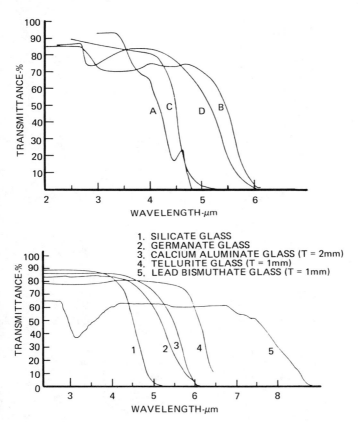

Figure 2.12    Transmittance of infrared transmitting glasses (see Table 2.19 for glass designations). (From Dumbaugh, 1981.)

surfaces are excluded.  Where necessary for the purposes of this comparison, the above curves have been calculated from data at other thicknesses (with material dispersion included in the surface reflectances).  Materials represented include:  UV grade synthetic fused silica (in particular, Amersil Inc.'s Suprasil), optical quality synthetic fused silica and BK-7. [Reprinted by permission of Melles Griot Co.; source:  Melles Griot Co. (1981).]

Table 2.19 Properties of IR Glasses[a]

A. Properties of IR-transmitting oxide glasses

| | Silicate[b] | Germanate[c] | Calcium[d] aluminate | Tellurite | Lead bismuthate |
|---|---|---|---|---|---|
| | 9753 | 9754 | | | |
| Expansion coefficient (25–300°C) × $10^7$/°C | 59.5 | 62 | 97 (20–500°C) | 130–200 | 100–150 |
| Annealing point (°C) | 832 | 735 | 800 | 250–400 | 250–300 |
| Density (g/cm$^3$) | 2.8 | 3.6 | 3.1 | 5–7 | ~7 |
| Refractive index (589.3 nm) | 1.605 | 1.664 | 1.676 | 1.8–2.3 | ~2.5 |
| Knoop hardness (100 g) | 595 | 512 | ~525 | ~200 | <200 |
| Weatherability | Excellent | Fair | Poor | Poor | Poor |
| Fundamental absorption of network former oxygen bond — Wave number (cm$^{-1}$) | 1190–870 | 960–820 | 847 | 770 | <333 |
| Fundamental absorption of network former oxygen bond — Wavelength (μm) | 8.4–11.5 | 10.4–12.2 | 11.8 | 13.0 | >30 |

B. Comparison of silicate and germanate glasses

| | A | B | C | D |
|---|---|---|---|---|
| Cation oxide % $\{$ SiO$_2$ | 100 | – | 29.1 | – |
| GeO$_2$ | – | 100 | – | 29.1 |
| Al$_2$O$_3$ | – | – | 42.3 | 42.3 |
| CaO | – | – | 28.6 | 28.6 |
| Expansion coefficient × 10$^7$/°C, (25–300°C) | 5.5 | 76.3 | 59.5 | 63.6 |
| Annealing point (°C) | 1180 | 541 | 832 | 758 |
| Density (g/cm$^3$) | 2.2 | 3.65 | 2.80 | 3.35 |
| Knoop hardness (100 g) | 515 | – | 595 | 512 |
| Young's modulus, × 10$^{-3}$ kg/mm$^2$ | 7.44 | 4.42 | 10.05 | 9.56 |
| Refractive index (589.3 nm) | 1.458 | 1.607 | 1.605 | 1.660 |
| Abbe number | 67.8 | 41.6 | 55.0 | 46.4 |

[a]Source: Dumbaugh (1981).
[b]Corning Code 9753 glass.
[c]Corning Code 9754 glass.
[d]Barr and Stroud BS39B glass.

Transmittance curves for a variety of glasses used in the IR are presented in Fig. 2.12. Typical properties of the various IR glasses are displayed in Tables 2.19A and 2.19B.

For applications in the 8—14 μm range, crystalline compounds such as ZnS and ZnSe possess the appropriate cutoff. Beyond 14 μm, crystalline gallium arsenide and germanium metal are among the materials used. Many of the halide salts are typically employed in long-wave infrared applications although these materials have little resistance to environmental moisture and are soft and easily damaged.

A great deal of work has been performed to develop glasses based on germanium, calcium aluminates, tellurites, and lead bismuthates for IR application.

More recently a series of fluoride glasses has been developed with extremely low absorption in response to the demands of laser optics and fiber optics. These fluoride-based glasses are transparent to about 8 μm. The relatively inexpensive new fluoride glasses are synthesized from the $ZrF_4$—$BaF_2$ and the $HfB_4$—$BaF_2$ systems (Drexhage, 1981).

## 2.15 OPTICAL GLASS FIBERS

The advent of the long distance transmission of information by optical signals became possible with the development of glass fibers with exceptionally low absorption coefficients. Conventional glass has 500—1000 dB/km attenuation. Such fibers are adequate for 15—18 m of line. In contrast, special low loss fused silica has a 1 dB/km loss.

Ten times the common logarithm of the ratio of the initial to the terminal energy flow is an index of the attenuation of a plane wave in decibels (dB):

$$A_{dB} = 10 \log_{10} I_0/I$$

where $I_0$ is the initial energy and $I$ the terminal energy. Thus, for a 1 dB attenuation, we have

$$1 = 10 \log_{10} I_0/I$$
$$\log_{10}(I_0/I) = 0.1 \quad \text{or}$$
$$(I_0/I) = 1.259 \quad \text{or}$$
$$I = 0.749 I_0$$

Silica fibers with attenuation of only 0.2 dB/km have also been fabricated. This fiber loses only half its initial optical power over a distance of 15 km.

Characteristic absorption coefficients for a dielectric material are shown schematically in Fig. 2.13. For communications, materials are sought where the minimum absorption coefficient coincides with the transmitting wavelength of interest.

For short range optical links cheaper plastic fibers may be used. These are made from a core of polystyrene or polymethyl metacrylate clad with a polymer of lower refractive index. The best of these have a measured attenuation of 200 dB/km at 590 nm and 300 dB/km at 650—670 nm.

Zinc chloride glass has been demonstrated to have attenuation of $10^{-3}$ dB/km in the 3.5—4.0 μm region. Some halide compounds are exceptional such as thallium bromide at $10^{-5}$ dB/km in the 4—5.5 μm bandpass.

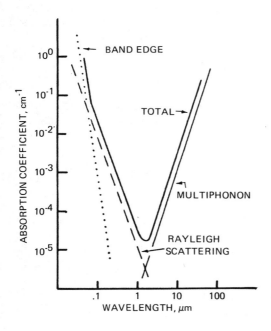

Figure 2.13   Absorption coefficient versus wavelength for a typical dielectric material (from Drexhage, 1981).

Optical fibers are clad with a lower index of refraction material so that the light being transmitted is totally reflected internally.

The first low loss optical fiber was reported by Corning. To attain such low losses, transition metal impurity contents must be held in the <50 ppb range and hydroxyl ion content must be held to <50 ppb as well. These fibers were made by the Corning doped deposited silica (DDS) technology. This process is described in Chap. 5.

A recent demonstration of a 101 km single mode fiber made by this technique showed an average loss of <0.4 dB/km at 1300 nm.

## REFERENCES

M. G. Drexhage, Fluoride glasses for visible to mid-IR guided wave optics, In *Advances in Ceramics*, American Ceramic Society, Columbus, OH, 1981, Vol. 2.

W. H. Dumbaugh, Infrared Transmitting Germanate Glass, *Emerging Optical Materials*, SPIE, Vol. 297, Bellingham, WA, 1981.

P. W. McMillan, *Glass Ceramics*, Academic Press, New York, 1979.

Melles Griot Co., *Optics Guide 2*, Irvine, CA, 1981.

G. W. Morey, *The Properties of Glass*, Reinhold Publishing Corp., New York, 1954.

Ohara Optical Glass, Inc. *General Catalogue*, Wachung, NJ, 1982.

Schott Glass Technologies, Inc. *Optical Glass*, Duryea, PA, 1982.

F. V. Tooley, *Handbook of Glass Manufacture*, Ogden Publishing Co., New York, 1960.

# 3
# CRYSTALLINE AND TWO-PHASE MATERIALS

## 3.1 INTRODUCTION

In the context of this book, ceramics are defined as members of that class of materials which are inorganic compounds of metal and nonmetal ions. These materials may be glasses, crystalline substances or combinations of glasses and crystals. The most widely used materials for optical refractors are glasses of various compositions, as discussed in Chap. 2.

Most of the development work performed over the past century for improved materials for application to optical elements has been in the field of the glass sciences. Nevertheless, nonglasses are becoming increasingly important because of their optical, mechanical, chemical, and thermal properties which cannot be matched by glass in many instances. A simple example of such a property is the abrasion resistance of sapphire.

This chapter describes crystals, glass ceramics, and two-phase glasses which are important to the art and science of optical devices.

## 3.2 CRYSTALLINE MATERIALS

### Single Crystals

The crystalline state is characterized by position regularity of the atoms constituting the solid. The bonding between the atoms in a crystal can be ionic, covalent, or intermediate in nature. In an ionic bond, the positive and negative ions have noninterpenetrating electron shells and the most stable separation between the positively and negatively charged ions is relatively large. The classical example is the sodium chloride (NaCl) crystal composed of a regular array of alternating $Na^+$ and $Cl^-$ ions. In this cubic array each $Na^+$ ion is surrounded by six $Cl^-$ ions and each $Cl^-$ ion is surrounded by six $Na^+$ ions. Ionic crystals are generally transparent in the visible and near infrared and exhibit low electrical conductivity at low temperature which increases at high temperature.

Covalent bonding is characterized by a pair of shared electrons between two atoms. The covalent bond is strongly directional. For example, the carbon atom forms tetrahedral bonds as in the diamond structure. Covalent crystals have high hardness, are refractory with high melting points, and when pure, have low electrical conductivity at low temperature. Covalent crystals are formed from atoms not too close in electronic structure to the inert gas electronic configuration, such as C, Ge, Si, and Te.

Many crystals are neither wholly ionic nor wholly covalent but fall between the two types with intermediate properties.

Only certain geometries can be repeated to fill space. There are only 32 permissible arrangements of points around a central origin that will fill space. Only 14 different space lattices organized in 7 systems are needed to accommodate these 32 permissible arrangements. The 14 Bravais or space lattices are illustrated in Fig. 3.1.

Crystals are characterized by lattice parameters of unit cell dimensions and the angles between the unit cell edges, as indicated in Fig. 3.2 and Table 3.1. Generally, the optical properties of crystals are different for various orientations of the crystal with respect to an incoming light ray. However, crystals with cubic symmetry exhibit no birefringence, while noncubic crystals exhibit birefringence effects. In single crystals, the main effect is the splitting of a light ray unless the ray is in the direction of one of the axes of the crystal. In polycrystalline optical solids the birefringence gives rise to internal scattering effects which depend on the individual crystallite size and orientation, the degree of birefringence, and the wavelength of interest.

Examples of cubic crystals are NaCl, diamond and MgO. On the other hand $\alpha$-quartz ($SiO_2$) and sapphire ($Al_2O_3$) crystals are hexagonal. Crystalline silica is polymorphic, existing in various forms dependent on the exact process which occurred in its genesis. The various polymorphs of silica and the thermal routes to their formation are shown in Fig. 3.3.

Many natural crystals have been found and an excellent collection resides in the Museum of Natural History in New York City. Some of the more common halide crystals are rocksalt (NaCl), sylbite (KCl), and fluorite (CaF). These are all cubic crystals and are suitable, in principle, for infrared spectroscopic windows. However, the naturally occurring crystals are not sufficiently large for many commercial applications.

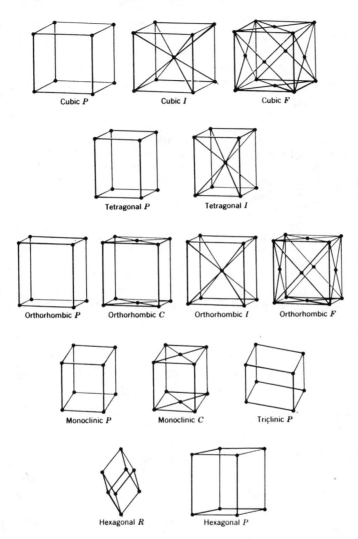

Cubic *P*        Cubic *I*        Cubic *F*

Tetragonal *P*        Tetragonal *I*

Orthorhombic *P*    Orthorhombic *C*    Orthorhombic *I*    Orthorhombic *F*

Monoclinic *P*        Monoclinic *C*        Triclinic *P*

Hexagonal *R*        Hexagonal *P*

**Figure 3.1**    Fourteen space (Bravais) lattices (from Kingery,
Bowen, and Uhlmann, 1976).  Reprinted with permission from W. D.
Kingery, H. K. Bowen, and D. R. Uhlmann, *Introduction to Ceram-*
*ics*, John Wiley and Sons, Inc., New York, 1976.

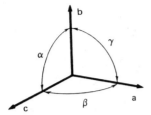

**Figure 3.2**  Lattice parameters of the unit cell (from Kingery, Bowen, and Uhlmann, 1976). Reprinted with permission form W. D. Kingery, H. K. Bowen, and D. R. Uhlmann, *Introduction to Ceramics*, John Wiley and Sons, Inc., New York, 1976.

**Table 3.1**   Unit Cell Parameters[a]

| System | Number of lattices | Lattice symbols | Restrictions on conventional cell axes and angles |
|---|---|---|---|
| Triclinic | 1 | P | $a \neq b \neq c$ <br> $\alpha \neq \beta \neq \gamma$ |
| Monoclinic | 2 | P,C | $a \neq b \neq c$ <br> $\alpha = \gamma = 90° \neq \beta$ |
| Orthorhombic | 4 | P,C,I,F | $a \neq b \neq c$ <br> $\alpha = \beta = \gamma = 90°$ |
| Tetragonal | 2 | P,I | $a = b \neq c$ <br> $\alpha = \beta = \gamma = 90°$ |
| Cubic | 3 | P or sc <br> I or bcc <br> F or fcc | $a = b = c$ <br> $\alpha = \beta = \gamma = 90°$ |
| Trigonal | 1 | R | $a = b = c$ <br> $\alpha = \beta = \gamma < 120°, \neq 90°$ |
| Hexagonal | 1 | P | $a = b \neq c$ <br> $\alpha = \beta = 90°$ <br> $\gamma = 120°$ |

[a]Reprinted with permission from W. D. Kingery, H. K. Bowen, and D. R. Uhlmann, *Introduction to Ceramics*, John Wiley and Sons, Inc., New York, 1976.

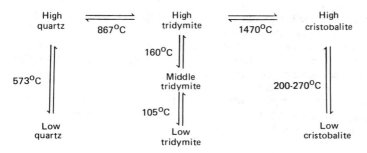

**Figure 3.3** Crystalline polymorphs of silica. Transformations of crystalline silica showing various transformations possible and the temperatures at which the transformations take place (from Kingery, Bowen, and Uhlmann, 1976). Reprinted with permission from W. D. Kingery, H. K. Bowen, and D. R. Uhlmann, *Introduction to Ceramics*, John Wiley and Sons, Inc., New York, 1976.

The first synthetic crystals were developed in the 1920s and became commercially available in the 1930s for IR spectroscopy. The crystals of major importance were found to be LiF, NaCl, CsBr, and KBr.

In the 1960s ruby ($Al_2O_3$ doped with 0.5% Cr) and YAG ($Y_3 Al_5 O_{12}$ + Nd) for 1.06 μm lasing crystals were grown by the Czochralski method. Good lasing rods up to 12 mm diam × 150 mm length were grown by this method.

The Czochralski technique, which was invented in 1917, is also known as crystal pulling. The molten material is contained in a suitable crucible and a seed crystal is slowly immersed into the liquid and then slowly withdrawn as it is rotated. The process is described in Chap. 4. This technique has been used also to grow large crystals of NaCl and KCl.

Another important technique for growing large transparent windows is the chemical vapor deposition (CVD) process. This process depends on the chemical reactions of gaseous species in a controlled environment where the thermal, pressure and flow conditions induce the formation of a solid body on a substrate. Among the most successful CVD-produced optical materials are the ZnS and ZnSe infrared transmitting polycrystalline ceramics.

Natural quartz crystals can be found. However, quartz crystals are grown synthetically for the communications industry, which

uses them on the order of 100 tons/yr. These quartz crystals
are grown in an isostatic pressure vessel held at 350°C/100 MPa
($\sim$14,500 psi). The vessel contains a solution of $SiO_2$ into which
single crystal quartz seeds are suspended. Crystals grown in this
manner have extremely good ultraviolet transmission, rising to near
100% at 300 nm with a sharp cutoff at 147 nm.

The optical transmitting ranges (bandpasses) for a variety of
crystalline materials (and fused quartz) are listed in Table 3.2.
Most of those listed are binary compounds.

### Complex Crystals

Crystalline materials with more complex chemistries exist in many
binary, ternary, and higher order phase equilibrium systems. For
example, spinel $MgO \cdot Al_2O_3$ occurs in the $MgO-Al_2O_3$ system illustrated in Fig. 3.4. The phase diagram can be considered as a map
which depicts the phases present in equilibrium at any composition
and temperature. Equilibrium implies that the components are in
a stable state and that no further changes will occur as long as the
conditions remain invariant. Each phase is saturated with all other
phases in contact, but not supersaturated.

Most published phase diagrams for ceramic materials are based
on an air environment at one atmosphere. The phase diagram boundaries can shift for other pressures or other compositions of the surrounding atmosphere. In Fig. 3.4 the single phase spinel exists
in a region approximately 50 M% $Al_2O_3$, 50 M% MgO. However, the
spinel phase at 1750°C can have a composition varying from 47 to
71 M% MgO.

Spinel has a cubic structure and therefore does not exhibit birefringence. It can be synthesized as a single crystal or as a polycrystalline body.

Another well-known system is the $SiO_2-Al_2O_3$ series of compositions. At the one end is found the optically important silica and
at the other, alumina which is optically significant as Lucalox® and
as sapphire. Again, a single phase exists, mullite, at about 60 M%
$Al_2O_3$. Mullite is not cubic and exhibits a moderate degree of birefringence. As the temperature of the mullite is increased, the material dissociates into alumina (solid) and a liquid phase at 1828°C
as shown in Fig. 3.5.

Many crystalline phases occur in the ternary system composed
of $SiO_2$, $Al_2O_3$, and MgO. A representation of the major phases is
provided in the ternary equilibrium diagram in Fig. 3.6 given in
weight percent (wt%) rather than in mole percent (M%) compositions. Any point in the diagram represents a fractional composition each component of which is given by the ratio d/A, where d is

**Table 3.2 Crystalline Optical Materials[a]**

Useful Transmission (Exceeding 10%) Regions of Materials for 2-mm Thickness

Ultraviolet — Visible — Near IR — Middle IR — Far IR — Extreme IR

.1 .2 .3 .4 .5 .6 .7 .8 .9 1 2 3 4 5 6 7 8 9 10 15 20 30 40 50 60 80 100 Microns

Fused Silica (SiO$_2$) .16 — 4
Fused Quartz (SiO$_2$) .18 — 4.2
Calcium Aluminate Glass (CaAl$_2$O$_4$) .35 — 5.5
Lithium Metaniobate (LiNbO$_3$) .4 — 5.5
Calcite (CaCO$_3$) .2 — 5.5
Titania (TiO$_2$) .43 — 6.2
Strontium Titanate (SrTiO$_3$) .39 — 6.8
Alumina (Al$_2$O$_3$) .2 — 7
Sapphire (Al$_2$O$_3$) .15 — 7.5
Lithium Fluoride (LiF) .12 — 8.5
Magnesium Fluoride (Polycrystalline) (MgF$_2$) .45 — 9
Yttrium Oxide (Y$_2$O$_3$) .26 — 9.2
Magnesium Oxide (Single Crystal) (MgO) .25 — 9.5
Magnesium Oxide (Polycrystalline) (MgO) .3 — 9.5
Magnesium Fluoride (Single Crystal) (MgF$_2$) .15 — 9.6
Calcium Fluoride (Polycrystalline) (CaF$_2$) .13 — 11.8
Calcium Fluoride (Single Crystal) (CaF$_2$) .13 — 12
Barium Fluoride/Calcium Fluoride (BaF$_2$/CaF$_2$) .75 — 12
Arsenic Trisulfide Glass (As$_2$S$_3$) — 13
Zinc Sulfide (ZnS) .6 — 14.5
Sodium Fluoride (NaF) .14 — 15
Barium Fluoride (BaF$_2$) .13 — 15
Silicon (Si) 1.2 — 15
Lead Fluoride (PbF$_2$) .29 — 15
Cadmium Sulfide (CdS) .55 — 16
Zinc Selenide (ZnSe) .48 — 22
Germanium (Ge) 1.8 — 23
Sodium Iodide (NaI) .25 — 25
Sodium Chloride (NaCl) .2 — 25
Potassium Chloride (KCl) .21 — 25
Silver Chloride (AgCl) .4 — 30
Thallium Chloride (TlCl) .42 — 30
Cadmium Telluride (CdTe) .9 — 31
Thallium Chloro Bromide (TlCl, Br) .4 — 35
Potassium Bromide (KBr) .2 — 38
Silver Bromide (AgBr) .45 — 40
Thallium Bromide (TlBr) .38 — 40
Potassium Iodide (KI) .25 — 47
Thallium Bromo Iodide (TlBr, I) .55 — 50
Cesium Bromide (CsBr) .2 — 55
Cesium Iodide (CsI) .25 — 70

[a]See footnote to Table 3.1.

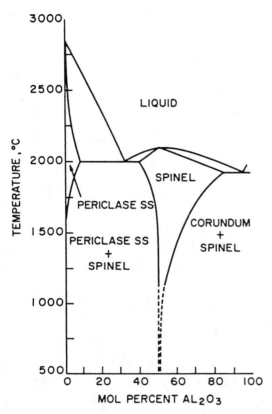

Figure 3.4    Phase diagram for the MgO–Al$_2$O$_3$ system; SS = solid
solution (from Levine, Robbins, and McMurdie, 1964).

the distance to the base of the unit edge of the equilateral triangle
with the component in question forming the apex of the triangle,
and A is the altitude of that triangle.

For example, mullite is shown as 0 wt% MgO, 73 wt% Al$_2$O$_3$, and
27 wt% SiO$_2$.  Cordierite (2MgO·2Al$_2$O$_3$·5SiO$_2$) is plotted at 14 wt%
MgO, 35 wt% Al$_2$O$_3$, and 51 wt% SiO$_2$.

### Diamond

Diamond is an exotic crystalline material.  The well-known diamond
crystal structure is shown in Fig. 3.7.  A single crystal natural

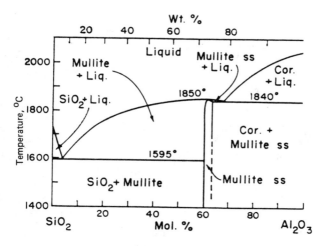

Figure 3.5   Binary system Al$_2$O$_3$–SiO$_2$; SS = solid solution (from Levine, Robbins, and McMurdie, 1964).

diamond has been used as an infrared window in an IR sensor for one of the probes to the planet Venus. Diamond does not have to be of gem quality to be optically useful. Suitable natural diamonds are available and, although expensive, these are much cheaper than gem quality stones. Diamond's outstanding optical property is its longwave cutoff, beyond 200 μm. Diamond is a cubic crystal and has no intrinsic birefringence. However, internal strains are manifested as birefringence (strain birefringence).

Diamonds are classified into four main types: Ia, Ib, IIa, and IIb. Each type has a characteristic transmission spectrum. Type I diamonds contain dissolved nitrogen. Type IIb diamonds contain dissolved boron and become semiconductors at high temperature. Most natural diamonds are type Ia. The index of refraction n of natural diamond in the visible varies between 2.40 and 2.46. Due to the high value of n, 17% of the incident light is reflected from the surface of the diamond. The density of cubic diamond is 3.15 g/cm$^3$ compared to 2.2 g/cm$^3$ for theoretically dense graphite (Chrenko and Strong, 1975). Diamond is, of course, the hardest substance known. Its Knoop hardness ranges from 8200 to 10,400 kg/mm$^2$. For comparison, crystalline quartz (SiO$_2$) is 750 kg/mm$^2$ Knoop hardness, while sapphire (Al$_2$O$_3$) is 2000 kg/mm$^2$. Diamond has a very high thermal conductivity, approximately six times that of copper at room temperature for a type IIa diamond.

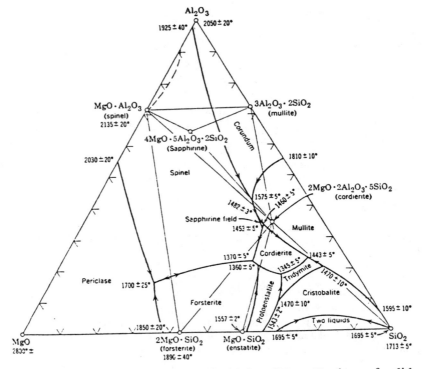

**Figure 3.6** Ternary system $MgO-Al_2O_3-SiO_2$. Regions of solid solution are not shown. (From Levine, Robbins, and McMurdie, 1964.)

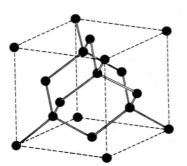

**Figure 3.7** Diamond structure.

Diamond can be synthesized at ultrahigh pressure. Catalytic conversion occurs in the neighborhood of 80 kbar (8000 MPa or $1.2 \times 10^6$ psi) and 1800°C as indicated in Fig. 3.8.

D. Drukker & ZN, Holland, fabricates type IIa diamonds for IR windows from 1/4 to 18 mm diameter. A large Drukker type IIa diamond went to Venus in the pressure shell of the Pioneer Large Probe and successfully resisted sulfuric acid clouds, near red heat, and 100 atm pressure.

Small Drukker diamonds will be on board the Galileo spacecraft when it goes to Jupiter. Others are circling the earth on a Nimbus

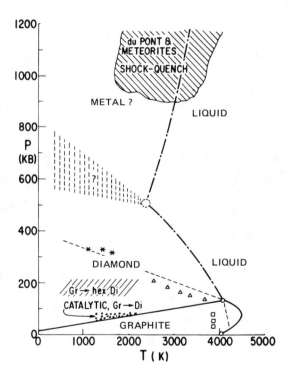

**Figure 3.8**  Carbon phase diagram: the direct transition (* and Δ points) lies on the metastable extension of the graphite melting line; (□) region of rapid spontaneous transformation of diamond to graphite (from Chrenko and Strong, 1975). Reprinted by permission of the General Electric Company, Corporate Research and Development.

weather satellite.  These windows have been essential for certain
types of space spectroscopy.


## 3.3  GLASS CERAMICS

Another class of materials which is useful for electromagnetic transmission is the glass ceramic material category.  As the name implies,
these substances are intermediate between the glassy state and the
crystalline state.  In fact, glass ceramics are composed of very fine
crystals dispersed in a glass matrix.  An important example is Zerodur (Schott), an ultralow expansion material used for telescope
mirror substrates.

A glass ceramic is made by incorporating suitable nucleating
agents such as $TiO_2$ in a glass of the proper composition for the
creation of a glass ceramic.  The glass (including incorporation of
the nucleating agent) is processed by conventional glassmaking
techniques.  When cooled, the object is amorphous.  However, upon
heat treating many small crystals nucleate almost simultaneously

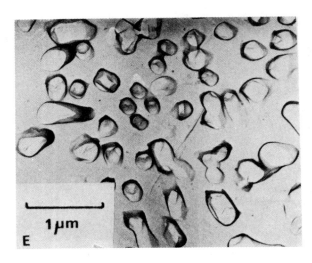

Figure 3.9   Microstructure of a glass ceramic $ZnO-Al_2O_3-SiO_2$
with $P_2O_5$ nucleant (from Partridge, Phillips, and Riley, 1973).

and grow around the nucleating agent in a controlled manner. The resultant structure is an amorphous matrix with an embedded crystalline phase. An example of the microstructure of a glass ceramic is shown in Fig. 3.9.

The product is transparent if there is a good index of refraction match between the glass and the crystal phases, if the crystallites are small relative to the wavelength of light, and if there is an absence of impurities and porosity.

Glass ceramics are useful because conventional (relatively inexpensive) glassmaking processes can be used to create a shape. The crystalline attributes are achieved by heat treating the glass blank so produced to develop a finely dispersed crystalline phase which can impart hardness, strength, low expansion coefficient, and chemical and thermal stability.

Such materials are used for mirrors and other optical elements where dimensional stability over a specified temperature range is vital.

Typical compositions for glass ceramics are given in Table 3.3, and a schematic heat treating schedule is shown in Fig. 3.10. A nucleation period is required followed by higher temperature crystal growth period to attain the optimum microstructure and phase composition. Ultra Low Expansion® (ULE) glass by Corning,

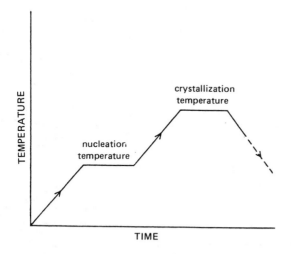

Figure 3.10    Heat treatment schedule for controlled crystallization of glass (from Partridge, Phillips, and Riley, 1973).

**Table 3.3**    Some Compositions of Glass Ceramics:  Influence of Various Nucleating Agents on Properties of Glass Ceramics

| | | | | | | | Composition | |
|---|---|---|---|---|---|---|---|---|
| | Main constituents (wt%) | | | | | | | |
| No. | $SiO_2$ | $Li_2O$ | $Na_2O$ | $K_2O$ | $ZnO$ | $Al_2O_3$ | $P_2O_5$ | $TiO_2$ |
| 1 | 83.0 | 17.0 | — | — | — | — | — | — |
| 2 | 80.6 | 16.7 | — | — | — | — | 2.7 | — |
| 3 | 83.0 | 17.0 | — | — | — | — | — | — |
| 4 | 72.0 | 20.0 | 2.0 | — | — | 6.0 | — | — |
| 5 | 71.2 | 19.8 | — | — | — | 6.0 | 3.0 | — |
| 6 | 81.0 | 12.5 | — | 2.5 | — | 4.0 | — | — |
| 7 | 78.5 | 12.1 | — | 2.5 | — | 3.9 | 3.0 | — |
| 8 | 61.0 | 9.2 | — | 2.0 | 27.8 | — | — | — |
| 9 | 59.0 | 9.0 | — | 2.0 | 27.3 | — | — | 2.7 |
| 10 | 57.7 | 8.8 | — | 1.9 | 26.6 | — | — | 5.0 |
| 11 | 59.0 | 9.0 | — | 2.0 | 27.3 | — | 2.7 | — |
| 12 | 59.0 | 9.0 | — | 2.0 | 27.3 | — | — | — |
| 13 | 63.5 | 5.7 | — | 4.2 | 5.3 | 18.9 | 2.4 | — |
| 14 | 62.5 | 5.6 | — | 4.2 | 5.3 | 18.5 | — | — |
| 15 | 61.0 | 5.6 | — | 4.1 | 5.2 | 18.2 | 2.4 | — |
| 16 | 61.7 | 5.7 | — | 4.2 | 5.3 | 18.5 | 2.4 | — |
| 17 | 62.5 | 5.7 | — | 4.2 | 5.3 | 18.5 | 2.4 | — |
| 18 | 62.5 | 5.7 | — | 4.2 | 5.3 | 18.9 | 2.4 | 1.3 |
| 19 | 63.4 | 5.7 | — | 4.2 | 5.3 | 18.9 | 2.4 | — |
| 20 | 63.5 | 5.7 | — | 4.2 | 5.3 | 18.9 | 2.4 | — |

[a]Source:  Partridge, Phillips, Riley (1973).

| Nucleating agents (wt%) | | | | | | Crystal size (μm) | Cross breaking strength (lb/in.[2]) |
|---|---|---|---|---|---|---|---|
| $MoO_3$ | $WO_3$ | $V_2O_5$ | Au | Ag | Pt | | |
| – | – | – | – | – | – | 15 | 3,400 |
| – | – | – | – | – | – | 1–3 | 59,000 |
| – | – | – | 0.027 | – | – | 15 | 5,000 |
| – | – | – | – | – | – | 10–20 | 9,300 |
| – | – | – | – | – | – | 5–10 | 30,800 |
| – | – | – | 0.027 | – | – | 5 | 38,000 |
| – | – | – | – | – | – | 5 | 40,000 |
| – | – | – | – | – | – | 20 | 300 |
| – | – | – | – | – | – | 20 | 300 |
| – | – | – | – | – | – | 5 | 300 |
| – | – | – | – | – | – | 5 | 25,500 |
| 2.7 | – | – | – | – | – | 5 | 36,000 |
| – | – | – | – | – | – | 10 | 12,500 |
| – | 3.9 | – | – | – | – | 10 | 15,900 |
| – | 3.5 | – | – | – | – | 10 | 24,200 |
| 2.2 | – | – | – | – | – | 10 | 28,200 |
| – | – | 1.4 | – | – | – | 10 | 23,600 |
| – | – | – | – | – | – | 10 | 16,900 |
| – | – | – | – | 0.1 | – | 15 | 4,300 |
| – | – | – | – | – | 0.001 | 15 | 5,700 |

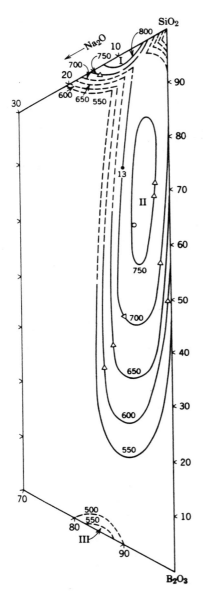

**Figure 3.11** Phase diagram for $SiO_2$–$B_2O_3$–$Na_2O$ showing regions of immiscibility at various temperatures (from Kingery, Bowen, and Uhlmann, 1976). Reprinted with permission form W. D. Kingery, H. K. Bowen, and D. R. Uhlmann, *Introduction to Ceramics*, John Wiley and Sons, Inc., New York, 1976.

Cer-Vit® by Owens Illinois, and Zerodur® by Schott Glass Technologies are examples of glass ceramics used for mirror substrates.
   Zerodur contains 70%—80% by weight of a high-temperature
quartz crystal phase with a mean crystallite size of 50—55 nm
(0.050—0.055 µm).  This crystallite phase has a negative coefficient of expansion which balances the positive thermal expansion
of the glassy (vitreous) matrix.  The coefficient of thermal expansion of Zerodur is $\alpha_{20-300°C}$ = +0.05 × $10^{-6}/°C$ (Schott,
1982).


## 3.4    TWO-PHASE GLASSES—PYREX

Another type of two-phase material is exemplified by Pyrex, a
boro-silicate two-phase glass.  It exhibits phase separation on a
fine scale, <50 Å (0.0050 µm).  A high-durability (low-Na) Pyrex
composition is

| | |
|---|---|
| $SiO_2$ | 81M% |
| $B_2O_3$ | 13M% |
| $Na_2O$ | 4M% |
| $Al_2O_3$ | 2M% |

Fig. 3.11 is a portion of the $SiO_2-B_2O_3-Na_2O$ equilibrium phase
diagram.  The region marked II represents a region where two im-
miscible liquid phases exist in equilibrium at 750°C.  Upon cooling
from the melt temperature, these two phases solidify on a very fine
scale of dispersion producing the Pyrex-like product, which is useful in some optical applications.  Pyrex glass is noted particularly
for its low thermal expansion coefficient and its high thermal shock
resistance.


## REFERENCES

R. M. Chrenko and H. M. Strong, *Physical Properties of Diamond*,
   Report No. 75CRD089, General Electric Co., Schenectady, NY,
   1975.
W. D. Kingery, H. K. Bowen, and D. R. Uhlmann, *Introduction
   to Ceramics*, John Wiley and Sons, Inc., NY, 1976.
E. M. Levine, C. R. Robbins, and H. F. McMurdie, *Phase Dia-
   grams for Ceramists*, The American Ceramic Society, Columbus,
   OH, 1964.

P. W. McMillan, *Glass Ceramics*, Academic Press, New York, 1970.

G. Partridge, S. V. Phillips, and J. N. Riley, Characterization of crystal phases, morphology and crystallization processes in glass ceramics, *Trans. Br. Ceram. Soc.*, 72(6), 1973.

Schott Glass Technologies, Inc., *Zerodur Glass Ceramics*, Duryea, PA, 1982.

# 4
# ADDITIONAL PROPERTIES OF BULK INORGANIC OPTICAL MATERIALS

## 4.1 INTRODUCTION

Optical, thermophysical, and mechanical data for a variety of optical materials are provided in this chapter. These generic data may be used as design guides, but it is essential to obtain precise data from the manufacturer of the specific material of interest for reliable information. Properties of a given material can vary widely depending on impurity levels, processing variables, precise composition, microstructure, degree of crystallinity, and state of internal strain.

Mechanical strength is highly dependent on the details of the fabrication process. Therefore, any strength data provided herein are merely an indication of that property. The strength of a ceramic body is extremely sensitive to flaw size, population, and location within and on the surface. When performing structural analysis of ceramic optical elements, it is essential to exercise the principles of statistical fracture analysis.

Excellent collections of data are available in *The Infrared Handbook* (Wolfe and Zissis, 1978) and in the *American Institute of Physics Handbook* (Gray, 1963). In addition, manufacturers' catalogs provide much detailed physical property information for the vendors' products. For example, in the optical glass field, excellent data compilations are available from Schott Glass Technologies, Inc. (Schott, 1982), Ohara Optical Glass, Inc. (Ohara, 1982), and in the area of IR transmitting materials, Eastman Kodak's catalog on the Irtran materials (Eastman Kodak Company, 1971). Some of the critical properties of interest to the designer for a variety of optical materials are given in Tables 4.1A–4.1E.

77

**Table 4.1A**   Optical Properties of Candidate Laser Window Materials—Oxides Plus ZnSe and ZnS[a]

| Property | $\lambda$ ($\mu$m) | Sapphire $Al_2O_3$ | Fused silica $SiO_2$ | |
|---|---|---|---|---|
| High transmission spectral range ($\mu$m) | | 0.24—4 | 0.2—2.5 | |
| Index of refraction, n | 1.319 | $n_o$ = 1.7502 | 1.44634 | |
| | 3.8 | $n_o$ = 1.68368 | 1.39591 | |
| Thermal—optic coefficient $(\partial n/\partial T) \div 10^{-6}$ $K^{-1}$ @ $\sim$300K | 0.5876 0.5893 0.6328 0.6678 0.7679 1.15 3.39 10.6 | 1.36(o), 14.7(e) | 8.7 | |
| Effective optical absorption coefficient $\div 10^{-3}$ $cm^{-1}$ | 1.3 | 0.82 ± 0.16 | Suprasil W-1 GE 124 Heraus-Amersil    T-16 Ultrasil    T-17 Infrasil    T-12 Optosil    T-15 Homosil    T-08 Comm. Suprasil II Corning 7940 | 0.83 1.08  1.12 1.22 1.33 1.39 1.46 2.83 3.18 |
| | 3.8 | 28.5 | — | |
| Piezo-optic constants $\div 10^{-12}$ $Pa^{-1}$ | | | | |
| $q_{11}$ | 0.6328 0.6438 1.15 10.6 | -0.52 | | |
| $q_{12}$ | 0.6328 0.6438 | 0.08 | | |

| MgO | YAG $Y_3Al_5O_{12}$ | CVD ZnSe | CVD ZnS |
|---|---|---|---|
| 0.4−6 | 0.3−4 | 0.65−18 | 0.9−13 water clear 0.4−13 |
| 1.717714 | 1.8135 | 2.473 (1.15 μ) | 2.279 |
| 1.673380 | 1.7659 | | |
| 15.3 | | | |
| 11.6 ± 6.6 14.4 13.6 | 7.3 | 91.1, 107 | 63.5 |
| | | 59.7−70 53.4−62 52.0−61 | 46−49.8 43−45.9 41−46.3 |
| 8.1 ± 1.4 | 0.66 ± 0.08 | 7.38 ± 0.96 | 19.9 2.0 (water clear) |
| 9.4 | | 1.58 | 20.7 |
| | | −1.32, −1.44 | |
| | | −1.36 −1.39, −1.46 0.28, 0.17 | |

Table 4.1A    (Continued)

| Property | $\lambda$ ($\mu$m) | Sapphire $Al_2O_3$ | Fused silica $SiO_2$ |
|---|---|---|---|
| | 1.15 | | |
| | 10.6 | | |
| $q_{13}$ | 0.6428 | 0.13 | |
| $q_{14}$ | 0.6428 | -0.07 | |
| $q_{33}$ | 0.6428 | -0.41 | |
| $q_{44}$ | 0.6428 | -0.71 | |
| $q_1-q_{12}$ | 0.6328 | | |
| | 1.15 | | |
| | 10.6 | | |

[a]Source:   Fernelius, Graves, and Knecht (1981).

Table 4.1B   Optical Properties of Candidate Laser Window Materials—Fluorides[a]

| Property | $\lambda$ ($\mu$m) | $CaF_2$ |
|---|---|---|
| High transmission spectral range ($\mu$m) | | 0.25—7 |
| Index of refraction, n | 1.319 | 1.42712 |
| | 3.8 | 1.41147 |
| Thermal-optic coefficient $(\partial n/\partial T) \div 10^{-6}$ K1 @ $\sim$300 K | 0.4358 | |
| | 0.442 | |
| | 0.5561 | |
| | 0.5780 | |
| | 0.6328 | -1.8 to -1.1 |

| MgO | YAG $Y_3Al_5O_{12}$ | CVD ZnSe | CVD ZnS |
|---|---|---|---|
|  |  | 0.62 |  |
|  |  | 0.58, 0.51 |  |
|  |  |  |  |
|  |  | -1.60 |  |
|  |  | -1.98 |  |
|  |  | -1.97 |  |

| $BaF_2$ | $SrF_2$ | YLF $LiYF_4$ | ZBT fluoride glass |
|---|---|---|---|
| 0.22−9 | 0.3−9 | 0.22−5 | 0.4−5.5 |
| 1.46690 | 1.43161 | $n_o$ = 1.44655 | (1.53 in |
|  |  | $n_e$ = 1.46851 | visible) |
| 1.45768 | 1.42087 | $n_o$ = 1.42167 |  |
|  |  | $n_e$ = 1.44759 |  |
|  |  | -0.54(o), |  |
|  |  | -2.44(e) |  |
|  |  | -3.0($\pi$), |  |
|  |  | -0.8($\sigma$) |  |
|  |  | -0.67(o), |  |
|  |  | -2.30(e) |  |
|  |  | -0.91(o), |  |
|  |  | -2.86(e) |  |
| -1.67 to -1.63 | -1.25 to -1.20 | -5.6 |  |

**Table 4.1B** (Continued)

| Property | $\lambda$ ($\mu$m) | $CaF_2$ |
|---|---|---|
| | 1.15 | -1.34 to -1.18 |
| | 3.39 | -1.28 to -1.14 |
| Effective optical absorprion coefficient, $\beta_{eff} \div 10^{-3}$ cm$^{-1}$ | 1.3 | 0.39 (forged) |
| | | 0.71 (single crystal) |
| | 3.8 | |
| Piezo-optic constants $\div 10^{-12}$ Pa$^{-1}$ | | |
| $q_{11}$ | 0.6328 | -0.38 ± 0.03 |
| | 1.15 | -0.40 ± 0.06 |
| | 3.39 | -0.52 ± 0.11 |
| $q_{12}$ | 0.6328 | 1.08 ± 0.03 |
| | 1.15 | 1.09 ± 0.06 |
| | 3.39 | 1.00 ± 0.11 |
| $q_{44}$ | 0.6328 | 0.71 ± 0.01 |
| | 1.15 | 0.72 ± 0.01 |
| | 3.39 | 0.87 ± 0.06 |
| $q_{11} - q_{12}$ | 0.6328 | -1.46 ± 0.01 |
| | 1.15 | -1.49 ± 0.02 |
| | 3.39 | -1.51 ± 0.03 |
| Stress—optical coefficient $C = -\frac{n_o^3}{2}(q_{11} - q_{12})$ $\div 10^{-12}$ Pa$^{-1}$ | 0.6328 | 2.15 |
| | 1.15 | 2.16 |
| | 3.39 | 2.14 |

[a]Source: Fernelius, Graves, Knecht (1981).

| BaF$_2$ | SrF$_2$ | YLF LiYF$_4$ | ZBT fluoride glass |
|---|---|---|---|
| -1.71 to -1.66 | -1.28 to -1.27 | | |
| -1.68 to -1.62 | -1.30 to -1.26 | | |
| 1.29 | 1.56 | 0.344 | 3.27 ± 0.20 |
| | | 2.12 to 6.3 | 2 |
| | | | |
| -0.99 ± 0.03 | -0.64 ± 0.04 | | |
| -0.91 ± 0.07 | -0.63 ± 0.05 | | |
| -0.75 ± 0.07 | -0.83 ± 0.09 | | |
| 2.07 ± 0.04 | 1.45 ± 0.04 | | |
| 2.13 ± 0.07 | 1.50 ± 0.06 | | |
| 2.11 ± 0.05 | 1.23 ± 0.07 | | |
| 0.95 ± 0.01 | 0.60 ± 0.01 | | |
| 0.95 ± 0.01 | 0.62 ± 0.02 | | |
| 0.99 ± 0.07 | 0.72 ± 0.04 | | |
| -3.06 ± 0.01 | -2.08 ± 0.01 | | |
| -3.03 ± 0.02 | -2.13 ± 0.04 | | |
| -2.91 ± 0.08 | -2.05 ± 0.06 | | |
| 4.89 | 3.08 | | |
| 4.79 | 3.13 | | |
| 4.52 | 2.95 | | |

**Table 4.1C**   Thermal Properties of Candidate Laser Window Materials—Oxides Plus ZnSe and ZnS[a]

| Property | Sapphire $Al_2O_3$ | Fused silica $SiO_2$ |
|---|---|---|
| Specific heat C (J/g K) | @ 289.9 K    0.757<br>297.6 K    0.744<br><br>$\alpha$-$Al_2O_3$   @ 300 K<br>0.782 | Quartz glass<br>@ 293 K         0.687<br>285.09 K   0.720<br>295.05 K   0.738<br><br>305.04 K   0.755<br><br>298.15 K   0.739<br><br>298 K         0.754 |
| Thermal con-<br>ductivity $\kappa$<br>(W/cm K) | Linde synthetic single<br>crystal heat flow<br>parallel to optic axis<br>@ 299.2 K    0.251<br>heat flow perpen-<br>dicular to optic axis<br>@ 296.2 K    0.230 | Silky fused vitreous<br>silica<br>@ 317.1 K    0.0151<br>Fused silica,<br>Homosil<br>@ 314.2 K    0.0118<br>Various fused<br>quartz samples<br>@ 293 K      0.0137<br>302 K      0.0142<br>298 K      0.0138 |
| Coefficient of<br>linear<br>expansion $\alpha$<br>$\div 10^{-6}$ $K^{-1}$ | 5.8<br>5.4 | Recommended<br>value<br>@ 293 K<br>polycrystalline<br>10.3<br>fused<br>0.49 to 0.55 |
| Density $\rho$<br>$(g/cm^3)$ | 3.98 | 2.20 |

[a]Source:   Fernelius, Graves, and Knecht (1981).

| MgO | YAG $Y_3Al_5O_{12}$ | | CVD−ZnSe | CVD−ZnS |
|---|---|---|---|---|
| Fused @ 291 K 0.914 single crystal, Norton Co. @ 270 K 0.863 single crystal, Norton with impurities @ 273 K 0.883 323 K 0.966 99.9 MgO, fused @ 298 K 0.924 | @ 200 K 300 K | 0.456 0.626 | 0.339 | 0.469 |
| Single crystal @ 303.2 K 0.544 Bell Telephone Labs = 3.21 g/cm³ @ 315.4 K 0.359 318.5 K 0.361 | @ 200 K 300 K | 0.21 0.13 | 0.18 crystalline cubic 0.121 | 0.17 |
| @ 293 K +10.5 | @ 200 K 300 K 310 K | 5.8 7.5 6.9 | 7.57 7.2 cubic 7. | 7.85 |
| 3.58 | 4.55 | | 5.27 5.265 cubic 5.651 | 4.08 |

**Table 4.1D** Thermal Properties of Candidate Laser Window Materials—Fluorides [a]

| Property | CaF$_2$ | BaF$_2$ | SrF$_2$ | YLF LiYF$_4$ | ZBT fluoride glass |
|---|---|---|---|---|---|
| Specific heat $C$ (J/gK) | Large natural fluorite crystal @296.5 K  0.858 | @273.9 K  0.404 293.1 K  0.403 300.7 K  0.405 | Impurities 0.001–0.01 Ca, K, Cu, Fe, and Mg @290 K  0.554 295 K  0.556 300 K  0.559 | 0.79 | @45°C 0.511 |

| | Harshaw single crystal | Optovac, cubic isotropic crystal | 92% theoretical density | |
|---|---|---|---|---|
| Thermal conductivity κ (W/cm K) | Harshaw single crystal @318.7 K  0.0912 <br> Optovac single crystal @289  0.0985 <br> 309  0.0960 | Optovac, cubic isotropic crystal @311 K  0.0771 <br> Optovac crystal @284 K  0.0117 <br> 305 K  0.109 | 92% theoretical density @298 K  0.0142 <br> @300 K  0.083 | 0.06 |
| Coefficient of linear expansion $\alpha \div 10^{-6}$ K$^{-1}$ | 19.1 to 24 | 18.4 to 19.8 | 18.4 to 21.2 | a-axis  30–60°C <br> 13  4.3 <br> c-axis  250–270°C <br> 8  13.8 |
| Density ρ (g/cm³) | 3.179 | 4.83 to 4.90 | 4.18 to 4.278 | 3.99  4.8 |

[a] Source: Fernelius, Graves, and Knecht (1981).

Table 4.1E  Mechanical Properties of Prospective Laser Window Materials[a]

| Mechanical property | Al₂O₃ (single crystal) | MgO | SiO₂ (fused silica) | CaF₂ | SrF₂ | LiYF₄ | ZnSe | ZnS |
|---|---|---|---|---|---|---|---|---|
| **Flexural strength** | | | | | | | | |
| MPa | 480–1200 | 300 | 110 | 62–90 | 90 | 35 | 28–62 | 96–110 |
| ksi | 70–175 | 43 | 16 | 9–13 | 13 | 5 | 4–9 | 14–16 |
| **Compressive strength** | | | | | | | | |
| MPa | 2755 | | 750–1375 | | | | | |
| ksi | 400 | | 100–200 | | | | | |
| **Elastic modulus** | | | | | | | | |
| GPa | 400 | 240–345 | 72 | 83–110 | 84 | 7.5 | 69 | 76 |
| ×10⁶ psi | 58 | 35–50 | 10.5 | 9–16 | 12 | 11 | 10 | 11 |
| Poisson's ratio | 0.27 | 0.18 | 0.17 | 0.28 | 0.25 | 0.33 | 0.28 | 0.30 / 0.35 |

| | | | | | | | | |
|---|---|---|---|---|---|---|---|---|
| $K_{IC}$ MN m$^{-3/2}$ | 1–2 | | 0.4–0.7 | 0.4–0.6 | | | 0.6–1.3 | 0.6–1.4 |
| Fracture surface energy J(mm$^2$)$^{-1}$ | 0.2–5 | | | 0.5–0.8 | 0.4 | | 3–12 | 2–13 |
| Weibull parameters | | | | | | | | |
|   Shape-m value | 5–15 | | 4–5 | 3 | 3 | | 6–9 | 4–9 |
|   Scale (SI units) | 30–50 | | 6–7 | 5 | 4 | | 23–27 | 15–29 |
| Knoop hardness (100 g load) kg mm$^{-2}$ | 3000 | 800 | 575 | 200 | 150 | | 105 | 230 |
| Crystalline structure | Hexagonal | Cubic | Amorphous | Cubic | Cubic | Tetragonal | Cubic | Hexagonal |

[a]Source: Fernelius, Graves, and Knecht (1981).

## 4.2    OPTICAL PROPERTIES

Many documents display the internal or external transmittance versus wavelength data for materials of interest.  However, a single
such plot will not provide the absorption coefficient, $\alpha$, which is
needed for practical design.  In fact, a transmittance curve without
a thickness parameter stated is of minimal use.  From Bouguer's law
(see Sec. 1.3) and the Fresnel relations, the measurement of transmittance of two specimens of unequal thickness will provide the information needed to make a calculation of the absorption coefficient
from the following equation:

$$\alpha = \frac{-\ln(T_2/T_1)}{t_2 - t_1}$$

where T is the measured external transmittance and t is the corresponding thickness.

## 4.3    RESTSTRAHLEN REFLECTION

Another interesting property of crystalline materials is the reststrahlen reflection.  It peaks at a characteristic wavelength asso-

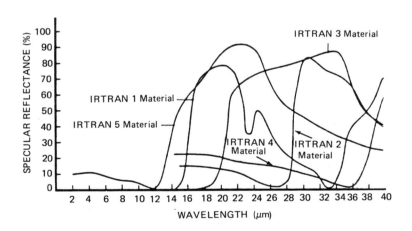

Figure 4.1    Reststrahlen reflection for Eastman Kodak Irtran mate-
rials.  Source:  Eastman Kodak Co. (1971).  Reprinted from Kodak
Publication No. U-72, *Kodak Irtran Infrared Optical Materials*,
Eastman Kodak Company, Rochester, NY, 1971.

ciated with the fundamental resonant vibration of the lattice.  In
approximate terms, the reststrahlen reflection occurs at a wavelength about twice that of the cutoff wavelength.  At resonant frequency, dielectric crystals behave as metallic reflectors.  In addition, at the resonant frequency, reflection increases with increasing angle of incidence.  Typical reststrahlen curves are shown in
Fig. 4.1 for the Eastman Kodak Irtran materials (Eastman Kodak
Company, 1971).

## 4.4  GLASSES

There are literally hundreds of glass compositions available from
the various optical glass suppliers.  Each firm publishes excellent
and comprehensive catalogs of data for its products.  Schott, Ohara,
Hoya, Corning, and Bausch and Lomb are among these suppliers.
    For illustrative purposes, data sheets for Schott glasses are
shown in Figs. 4.2 and 4.3.  Many of the definitions of these properties have already been reviewed in Chap. 2.

## 4.5  AMORPHOUS SILICA

There is some confusion in terminology concerning quartz.  Quartz
is a crystalline material.  The amorphous form of silica is known as
fused quartz or sometimes fused silica.  The fused material is available in a variety of grades from various manufacturers.  There are
at least two distinct processes, thermal fusion of natural crystalline
quartz (or natural quartz) and synthesis from high purity precursor
silicon tetrachloride.  The two processes yield products with distinctly different impurity content, about 20 ppm for the fused natural mineral, and 1 ppm for the $SiCl_4$ derived product.  Fused
quartz and fused silica are discussed further in Sec. 5.3.
    As a consequence of the high purity, the synthetic fused silica
cuts on at shorter wavelength in the UV and cuts off at a longer
wavelength in the IR.  Various grades of material are available from
major U.S. manufacturers such as Amersil, Inc., Dynasil, Inc.,
Corning Glass Works, and General Electric Co.  Typical transmission data for these materials have been shown in Fig. 2.11.
    Fused silica is generally an excellent optical material possessing
a very low coefficient of expansion and therefore very resistant to
thermal shock fracture and also having a high transmittance from
0.2 to 3 μm.  Representative properties of fused silica are given in
Table 4.1.

# BK 7 – 517 642

| | | |
|---|---|---|
| $n_d$ = 1.51680 | $v_d$ = 64.17 | $n_F - n_C$ = 0.008054 |
| $n_e$ = 1.51872 | $v_e$ = 63.96 | $n_{F'} - n_{C'}$ = 0.008110 |

### Refractive Indices

| | λ [nm] | |
|---|---|---|
| $n_{2325.4}$ | 2325.4 | 1.48929 |
| $n_{1970.1}$ | 1970.1 | 1.49500 |
| $n_{1529.6}$ | 1529.6 | 1.50094 |
| $n_{1060.0}$ | 1060.0 | 1.50669 |
| $n_t$ | 1014.0 | 1.50731 |
| $n_s$ | 852.1 | 1.50981 |
| $n_r$ | 706.5 | 1.51289 |
| $n_C$ | 656.3 | 1.51432 |
| $n_{C'}$ | 643.8 | 1.51472 |
| $n_{632.8}$ | 632.8 | 1.51509 |
| $n_D$ | 589.3 | 1.51673 |
| $n_d$ | 587.6 | 1.51680 |
| $n_e$ | 546.1 | 1.51872 |
| $n_F$ | 486.1 | 1.52238 |
| $n_{F'}$ | 480.0 | 1.52283 |
| $n_g$ | 435.8 | 1.52669 |
| $n_h$ | 404.7 | 1.53024 |
| $n_i$ | 365.0 | 1.53626 |
| | | |
| | | |
| | | |

### Constants of Dispersion Formula

| | |
|---|---|
| $A_0$ | 2.2718929 |
| $A_1$ | $-1.0108077 \cdot 10^{-2}$ |
| $A_2$ | $1.0592509 \cdot 10^{-2}$ |
| $A_3$ | $2.0816965 \cdot 10^{-4}$ |
| $A_4$ | $-7.6472538 \cdot 10^{-6}$ |
| $A_5$ | $4.9240991 \cdot 10^{-7}$ |
| | |
| | |

### Relative Partial Dispersion

| | |
|---|---|
| $P_{s,t}$ | 0.3097 |
| $P_{C,s}$ | 0.5607 |
| $P_{d,C}$ | 0.3075 |
| $P_{e,d}$ | 0.2386 |
| $P_{g,F}$ | 0.5350 |
| $P_{i,h}$ | 0.7478 |
| | |
| $P_{s,t}$ | 0.3075 |
| $P_{C',s}$ | 0.6058 |
| $P_{d,C'}$ | 0.2565 |
| $P_{e,d}$ | 0.2370 |
| $P_{g,F}$ | 0.4755 |
| $P_{i,h}$ | 0.7427 |

### Other Properties

| | |
|---|---|
| $\alpha_{-30/+70°C}$ [$10^{-6}$/K] | 7.1 |
| $\alpha_{20/300°C}$ [$10^{-6}$/K] | 8.3 |
| $T_g$ [°C] | 559 |
| $T_{10^{7.6}}$ [°C] | 719 |
| $c_p$ [J/g · K] | 0.858 |
| $\lambda$ [W/m · K] | 1.114 |
| | |
| $\rho$ [g/cm³] | 2.51 |
| $E$ [$10^3$ N/mm²] | 81 |
| $\mu$ | 0.208 |
| HK | 520 |
| | |
| B | 0 |
| CR | 2 |
| FR | 0 |
| SR | 1 |
| AR | 2.0 |

### Internal Transmittance $T_i$

| λ [nm] | $T_i$ (5 mm) | $T_i$ (25 mm) |
|---|---|---|
| 2325.4 | 0.89 | 0.57 |
| 1970.1 | 0.968 | 0.85 |
| 1529.6 | 0.997 | 0.985 |
| 1060.0 | 0.999 | 0.998 |
| 700 | 0.999 | 0.998 |
| 660 | 0.999 | 0.997 |
| 620 | 0.999 | 0.997 |
| 580 | 0.999 | 0.996 |
| 546.1 | 0.999 | 0.996 |
| 500 | 0.999 | 0.996 |
| 460 | 0.999 | 0.994 |
| 435.8 | 0.999 | 0.994 |
| 420 | 0.998 | 0.993 |
| 404.7 | 0.998 | 0.993 |
| 400 | 0.998 | 0.991 |
| 390 | 0.998 | 0.989 |
| 380 | 0.996 | 0.980 |
| 370 | 0.995 | 0.974 |
| 365.0 | 0.994 | 0.969 |
| 350 | 0.986 | 0.93 |
| 334.1 | 0.950 | 0.77 |
| 320 | 0.81 | 0.35 |
| 310 | 0.59 | 0.07 |
| 300 | 0.26 | |
| 290 | | |
| 280 | | |

### Remarks

| | |
|---|---|
| | |
| | |

### Deviation of Relative Partial Dispersions ΔP from the "Normal Line"

| | |
|---|---|
| $\Delta P_{C,t}$ | 0.0210 |
| $\Delta P_{C,s}$ | 0.0083 |
| $\Delta P_{F,e}$ | −0.0009 |
| $\Delta P_{g,F}$ | −0.0008 |
| $\Delta P_{i,g}$ | 0.0029 |
| | |

### Temperature Coefficients of Refractive Index

| | Δn/ΔT$_{relative}$ [$10^{-6}$/K] | | | | | Δn/ΔT$_{absolute}$ [$10^{-6}$/K] | | | | |
|---|---|---|---|---|---|---|---|---|---|---|
| [°C] | 1060.0 | s | C' | e | g | 1060.0 | s | C' | e | g |
| −40/−20 | 2.2 | 2.3 | 2.5 | 2.7 | 3.1 | 0.2 | 0.3 | 0.4 | 0.6 | 1.0 |
| −20/ 0 | 2.2 | 2.3 | 2.6 | 2.8 | 3.3 | 0.5 | 0.6 | 0.8 | 1.0 | 1.5 |
| 0/+20 | 2.3 | 2.4 | 2.7 | 2.8 | 3.4 | 0.9 | 1.0 | 1.2 | 1.3 | 1.9 |
| +20/+40 | 2.4 | 2.5 | 2.8 | 3.0 | 3.6 | 1.2 | 1.3 | 1.5 | 1.7 | 2.3 |
| +40/+60 | 2.5 | 2.6 | 2.9 | 3.1 | 3.8 | 1.3 | 1.4 | 1.7 | 1.9 | 2.6 |
| +60/+80 | 2.6 | 2.7 | 3.0 | 3.2 | 3.9 | 1.6 | 1.7 | 2.0 | 2.2 | 2.8 |

Figure 4.2    Properties of Schott Glass. Type BK7. Source: Schott Glass Technologies, Inc. (1982). Reprinted by permission of Schott Glass Technologies, Inc.

## Refractive indices

The refractive indices n are stated for max. 18 wavelengths between 365.0 nm and 2325.4 nm.

## Dispersion formula constants

With the aid of the constants $A_0 - A_5$, one can calculate the refractive indices for any required wavelength between 365.0 nm and by means of the dispersion formula
$n^2 = A_0 + A_1 \cdot \lambda^2 + A_2 \cdot \lambda^{-2} + A_3 \cdot \lambda^{-4} + A_4 \cdot \lambda^{-6} + A_5 \cdot \lambda^{-8}$.

## Relative partial dispersion

The relative partial dispersions $P_{xy}$ and $P'_{xy}$ for the wavelength x and y are calculated by means of the equations

$P_{xy} = \dfrac{n_x - n_y}{n_F - n_C}$ und $P'_{xy} = \dfrac{n_x - n_y}{n_{F'} - n_{C'}}$.

## Deviation of relative partial dispersion $\Delta P$ from the "normal line"

The term $\Delta P_{xy}$ gives a quantitative description of the deviation of dispersion behaviour from that of a "normal glass" (cf. "Secondary Spectrum").

## Internal transmittance $\tau_i$

The internal transmittance is an average value for a number of melts of the same type of glass, and is stated for the thicknesses 5 and 25 mm and wavelengths between 280 nm and 2,32.4 nm. Please use the circular slide rule when converting for other thicknesses (cf. Technical Information on Optical Glass TI, No. 3).

## Thermal coefficient of refraction

$\Delta n/\Delta T_{relative}$ at normal pressure 1.013 mbar
$\Delta n/\Delta T_{absolute}$, relative to vacuum.

## Other properties

$\alpha_{-30/+70}$ — Linear thermal expansion coefficient in the $-30\,°C$ to $+70\,°C$ temperature range, in $10^{-6}/K$

$\alpha_{20/300}$ — Linear thermal expansion coefficient in the $+20\,°C$ to $+300\,°C$ temperature range, in $10^{-6}/K$

$T_g$ — Transformation temperature, in °C

$T_{10^{7.6}}$ — Temperature of glass, in °C, at a viscosity of $1 \cdot 10^{7.6}$ dPa s

$c_p$ — Mean specific heat in J/g · K

$\lambda$ — Thermal conductivity, in W/m · K

$\rho$ — Density, in g/cm³

$E$ — Young's modulus, in $10^3$ N/mm²

$\mu$ — Poisson's ratio

HK — Knoop hardness

B — Bubble class
Total bubble volume per 100 cm³ of glass (inclusions are treated as bubbles). Expressed in bubble classes:
0 — few bubbles to 3 — numerous bubbles.

CR — Climatic resistance
Resistance to humidity, expressed in CR classes: 1 — high to 4 — slight.

FR — Stain resistance
Resistance to staining, expressed in FR classes: 0 — high to 5 — slight.

SR — Acid resistance
Resistance to acidic solutions, expressed in SR classes: 1 — high to 4 — slight and 51 – 53 (very slight).

AR — Alkaline resistance
Resistance to alkaline solutions, expressed in AR classes: 1 — high to 4 — slight with additional specification of surface change.

Figure 4.3   Explanation of symbols in Fig. 4.2.   Source: Schott Glass Technologies, Inc. (1982).   Reprinted by permission of Schott Glass Technologies, Inc.

## 4.6   CRYSTALLINE HALIDES

Considerable data have been developed on the halides. The halides are available as natural crystals and also can generally be synthesized as single crystals from molten salt baths by the Czochralski method of crystal growing. In the Czochralski technique, a seed crystal in contact with the molten salt is gradually raised while epitaxial growth of the crystal proceeds. The halides of interest for optical application are listed in Table 4.2.

Table 4.2    Properties of Halide Crystals[a]

| Material | Useful Transmission Range (microns) (by definition 60% for 10mm thickness) | IR Cutoff | Index of Refraction | Reflection Loss 2 Surfaces % | Solubility Water Grams/100 grams 25°C |
|---|---|---|---|---|---|
| NaCl (rocksalt) | | 16 | 1.495 (10μ) | 7.5 | 35.7 |
| KBr | | 25 | 1.525 (10μ) | 8.4 | 53.5 |
| KCl | | 20 | 1.457 (10 μ) | 6.8 | 34.7 |
| TlBr-TlI (KRS-5) | | 35 | 2.371 (10μ) | 28.4 | 0.05 |
| CsI | | 50 | 1.727 (20μ) | 13.6 | 44 |
| CaF₂ | | 9 | 1.400 (5μ) | 5.6 | insol. |
| BaF₂ | | 11.5 | 1.450 (5μ) | 6 | 0.004 |
| LiF | | 6.2 | 1.327 (5μ) | 4.4 | 0.014 |
| NaF | | 10.5 | 1.301 (5μ) | 3.6 | 4.22 |
| CdF₂ | | 9.8 | 1.576 (.59μ) | 5 | 4.35 |
| PbF₂ | | 10.9 | 1.708 (5μ) | 12 | .064 |
| MnF₂ | | 8.4 | 1.4 (6μ) | 4 | .0076 |
| SrF₂ | | 10.3 | 1.413 (5μ) | 6 | insol. |
| LaF₃ | | 9 | 1.60 (.55μ) | 11 | insol. |
| MgF₂ | | 7.0 | 1.377 (.58μ) | 4 | insol. |

| (Knoop) Hardness | Specific Gravity | Melting Point | Largest Diameter (inches) | Price ratio to NaCl | Special Characteristics | Applications | Material |
|---|---|---|---|---|---|---|---|
| 18.7 [100] 16.0 [110] | 2.164 | 801°C | 12 | 1 | Inexpensive, low reflection losses, relatively inert except to water. | 1 R Spectrometer windows cell windows. Laser windows. | NaCl (rocksalt) |
| 5.9–7.0 | 2.75 | 730°C | 8 | 2 | Transmits to 25m, very soluble in water, low index. | 1 R prisms, windows for long wavelength applications. Laser windows. | KBr |
| 7.2–9.3 | 1.99 | 776°C | 12 | 1 to 1.5 | Transmits 60% at 20m, low index, inert though water soluble. | Transmission to longer IR wavelengths than NaCl, otherwise similar. | KCl |
| 33.2–39.8 | 7.37 | 414°C | | | High index. Flows under pressure. | Excellent material for ATR prisms and FMIR plates | TlBr-TlI |
| (very soft) | 4.53 | 621°C | | | Soft, difficult to polish. | Prism and window material for long wavelength spectrophotometers. | CsI |
| 178 [100] 160 [110] | 3.18 | 1395°C | 10 | 4 | Low index. Takes excellent polish. High $\nu$ value. Inert to acids and bases except slightly sol. $H_2SO_4$. | Optical elements for U V vis. and I R regions. Prisms for I R, cell windows. Laser host. | $CaF_2$ |
| 87 [100] 76 [110] | 4.88 | 1355°C | 3 | 12 | Low index. Subject to fracture from thermal shock. Slightly sol. $H_2SO_4$. | Insol. material with longer transmission than $CaF_2$. | $BaF_2$ |
| 102–133 (600) | 2.64 | 844°C | 5 | 12 | U V Transmission is excellent in selected samples. | For U V Transmission to very short wavelengths. | LiF |
| (med. soft) | 2.79 | 1012°C | 6 | 24 | Low index. | Restrahlen Plate. | NaF |
| (med. soft) | 6.34 | 1100°C | 3 | | | Research interest. Semiconductor when treated. | $CdF_2$ |
| (med. soft) | 7.74 | 822°C | ¾ | | High density. | Research interest. | $PbF_2$ |
| (med. hard) | 3.98 | 930°C | 1.0 | | Low index. Hard. Anti-ferromagnetic. | Research interest. | $MnF_2$ |
| 154 [100] 140 [110] | 4.28 | 1450°C | 6 | 12 | Hard. slightly sol. $H_2SO_4$. | Laser Host. Similar to $CaF_2$. | $SrF_2$ |
| (med. hard) | 5.94 | 1493°C | ¾ | | Birefringent. | Laser Host and physics research. | $LaF_3$ |
| 415 | 3.18 | 1270°C | 6 | 12 | Low index, hard birefringent, high thermal & mechanical shock resistance. | Rugged I R and U V material for polarizers and optical elements. Low F-center formation and high laser-damage threshold. | $MgF_2$ |

Refractive index versus wavelength for some of these crystals is shown in Fig. 4.4, while dispersion versus wavelength is shown in Fig. 4.5. Transmission spectra and other optical and physical data are depicted in Figs. 4.6–4.16. The Irtran materials are IR-transparent hot-pressed polycrystalline bodies manufactured to exacting optical specifications.

Although not all the Eastman Kodak Irtran materials are halides, data for this popular class of IR transmitting materials are included in Figs. 4.17 and 4.18.

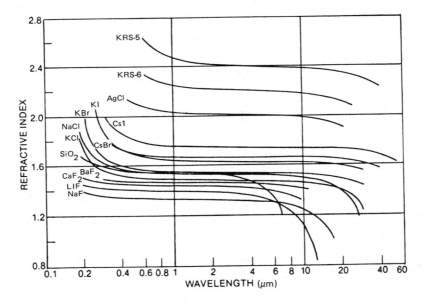

Figure 4.4    Refractive index for binary halides (from Harshaw, 1982).  Reprinted by permission of Harshaw/Filtrol Partnership.

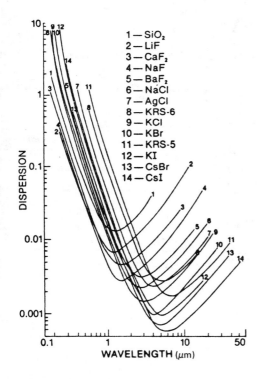

**Figure 4.5**   Dispersion versus wavelength for binary halides (from Harshaw, 1982). Reprinted by permission of Harshaw/Filtrol Partnership.

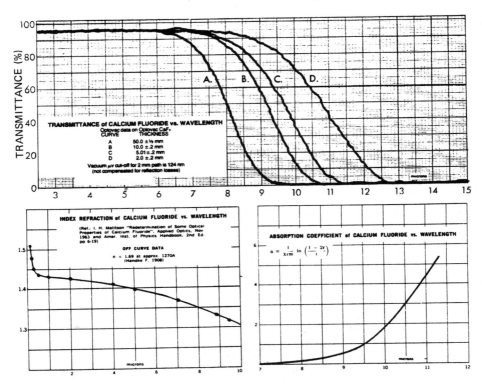

### General Physical Properties

| | | | |
|---|---|---|---|
| Structure | Cubic, cleaves 111 plane | Hardness | 178 Knoop (100); |
| Specific Gravity | 3.179 at 25°C | | 160 Knoop (110) |
| Dielectric Constant | 6.81 at 27°C | Elastic Constants | $C_{11} = 1.6420$ |
| Melting Point | 1395°C | $\times 10^{12}$ dyn/cm² | $C_{12} = 0.4398$ |
| Linear Thermal Expansion | $18.85 \times 10^{-6}/$ K | | $C_{44} = 0.3370$ |
| | at 300 K | Young's Modulus | $11 \times 10^6$ lb /in.² |
| Lattice Constant a | 5.45277A | Rigidity Modulus | $4.9 \times 10^6$ lb /in.² |
| Gruneisen Parameters | | Bulk Modulus | $12 \times 10^6$ lb /in.² |
| Specific Heat | 0.204 at 273 K | Rupture Modulus | $5.3 \times 10^3$ lb /in.² |
| Thermal Conductivity | 0.0232 cal/cm sec °C at 36°C | Apparent Elastic Limit | $5.3 \times 10^3$ lb /in.² |
| Water Solubility | Less than 0.0001 g /100 g H₂O at 23°C | | |

Figure 4.6    Properties of single crystal calcium fluoride (from Optovac, 1982).  Reprinted by permission of Optovac, Inc.

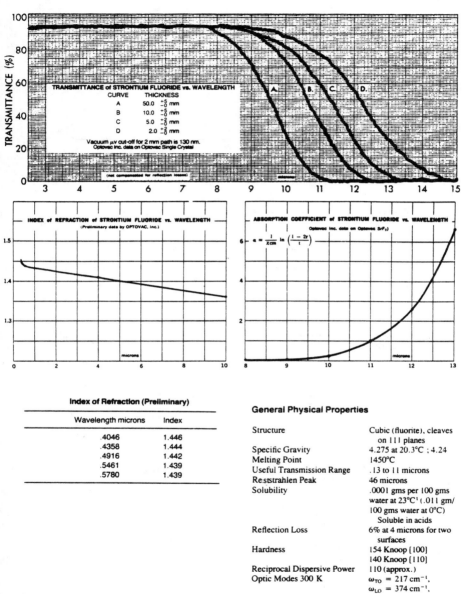

Figure 4.7   Properties of strontium fluoride (from Optovac, 1982).
Reprinted by permission of Optovac, Inc.

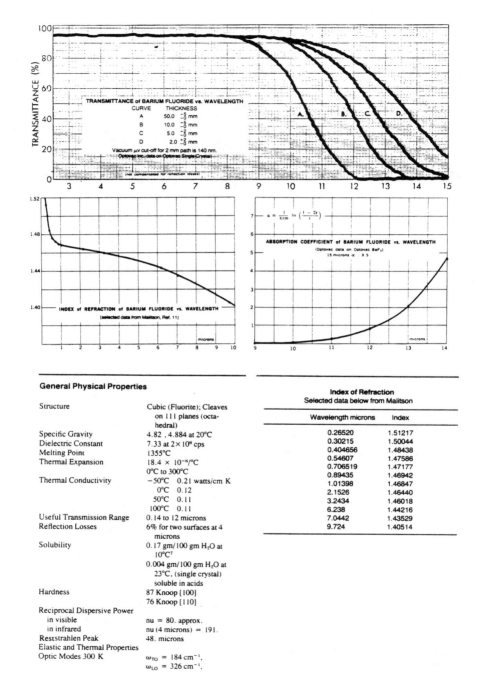

## General Physical Properties

| | |
|---|---|
| Structure | Cubic (Fluorite); Cleaves on 111 planes (octahedral) |
| Specific Gravity | 4.82 , 4.884 at 20°C |
| Dielectric Constant | 7.33 at $2 \times 10^6$ cps |
| Melting Point | 1355°C |
| Thermal Expansion | $18.4 \times 10^{-6}/°C$ 0°C to 300°C |
| Thermal Conductivity | −50°C  0.21 watts/cm K |
| | 0°C  0.12 |
| | 50°C  0.11 |
| | 100°C  0.11 |
| Useful Transmission Range | 0.14 to 12 microns |
| Reflection Losses | 6% for two surfaces at 4 microns |
| Solubility | 0.17 gm/100 gm $H_2O$ at 10°C[7] |
| | 0.004 gm/100 gm $H_2O$ at 23°C. (single crystal) soluble in acids |
| Hardness | 87 Knoop [100] |
| | 76 Knoop [110] |
| Reciprocal Dispersive Power | |
| in visible | nu = 80. approx. |
| in infrared | nu (4 microns) = 191. |
| Reststrahlen Peak | 48. microns |
| Elastic and Thermal Properties | |
| Optic Modes 300 K | $\omega_{TO} = 184\,cm^{-1}$, |
| | $\omega_{LO} = 326\,cm^{-1}$, |

### Index of Refraction
Selected data below from Malitson

| Wavelength microns | Index |
|---|---|
| 0.26520 | 1.51217 |
| 0.30215 | 1.50044 |
| 0.404656 | 1.48438 |
| 0.54607 | 1.47586 |
| 0.706519 | 1.47177 |
| 0.89435 | 1.46942 |
| 1.01398 | 1.46847 |
| 2.1526 | 1.46440 |
| 3.2434 | 1.46018 |
| 6.238 | 1.44216 |
| 7.0442 | 1.43529 |
| 9.724 | 1.40514 |

**Figure 4.8**   Properties of single crystal barium fluoride (from Optovac, 1982). Reprinted by permission of Optovac, Inc.

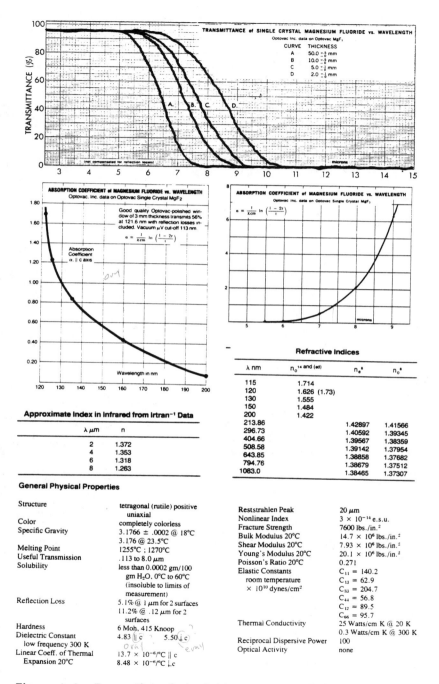

Figure 4.9   Properties of single crystal magnesium fluoride (from Optovac, 1982).   Reprinted by permission of Optovac, Inc.

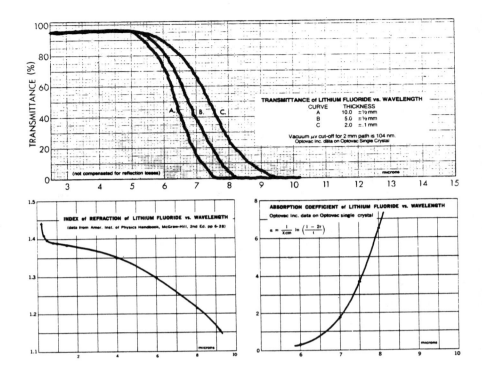

## General Physical Properties

| | |
|---|---|
| Structure | Cubic; cleaves on 100 plane (occasionally 110) |
| Specific Gravity | 2.639 at 25°C |
| | 2.640 at 22.2°C |
| Dielectric Constant | 9.00 at $10^2$—$10^7$ cps and 25°C |
| Melting Point | 848°C |
| Thermal Linear Expansion Coefficient | $37. \times 10^{-6}/°C$ 0° to 100°C |
| Thermal Conductivity | $2.7 \times 10^{-2}$ cal/cmsec°C at 41°C |
| Specific Heat | 0.373 at 10°C cal/gm/ K |
| Useful Transmission range | .104 to 7 microns |
| Reflection Loss | 4.4% for two surfaces at 4 microns |
| Solubility | 0.27 gm/100 gm $H_2O$ at 18°C[7] |
| | .014 gm/100 gm $H_2$) at 23°C |
| Hardness | Knoop 102 to 113 |
| Reststrahlen Peak | 25 microns |
| Elastic Constants $\times 10^{12}$ dynes/cm² | $C_{11} = .974$ |
| | $C_{12} = .404$ |
| | $C_{44} = .554$ |
| Elastic Properties $\times 10^6$ lbs/in² | Young`s Modulus, 9.4 |
| | Rigidity Modulus, 8.0 |
| | Bulk Modulus, 9.0 |

## Index of Refraction

| Wavelength microns | Index |
|---|---|
| 0.1250 | 1.585 |
| 0.1500 | 1.500 |
| 0.1750 | 1.463 |
| 0.1935 | 1.4450 |
| 0.254 | 1.41792 |
| 0.4861 | 1.39480 |
| 0.5000 | 1.39435 |
| 0.7000 | 1.39023 |
| 1.0 | 1.38711 |
| 2.0 | 1.37875 |
| 3.0 | 1.36660 |
| 4.0 | 1.34942 |
| 5.0 | 1.32661 |
| 6.0 | 1.29745 |

**Figure 4.10** Properties of single crystal lithium fluoride (from Optovac, 1982). Reprinted by permission of Optovac, Inc.

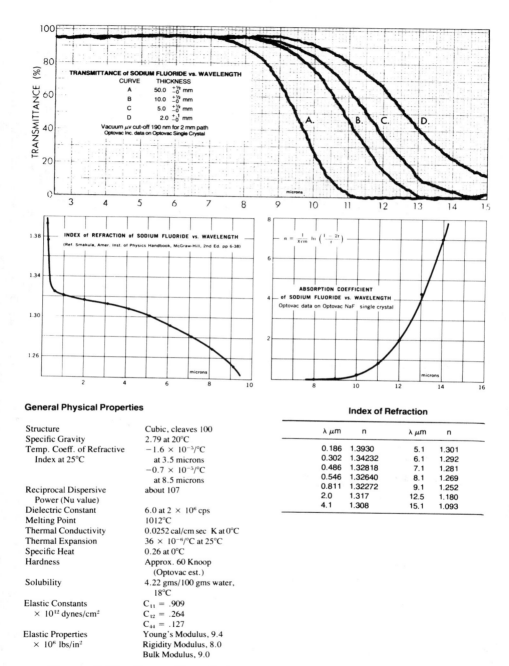

Figure 4.11   Properties of single crystal sodium fluoride (from Optovac, 1982).   Reprinted by permission of Optovac, Inc.

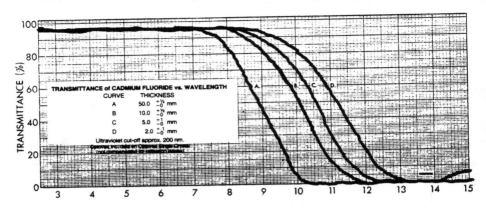

## General Physical Properties

| | |
|---|---|
| Structure | Cubic (fluorite), cleaves on 111 planes |
| Specific Gravity | 6.64 ; 6.34 at 24.5°C |
| Melting Point | 1100°C |
| Useful Transmission Range | .2 to 10 microns |
| Reflection Loss | approx. 6% for 2 surfaces |
| Solubility | 4.35 g/100 gm $H_2O$ at 25°C |
| Hardness | 3 Moh estimate |
| Index of Refraction | 1.576 for $n_D$ |
| Reciprocal Dispersive Power at $n_D$ | 61 |
| Nonlinear Refractive Index | |
| Elastic Constants 295 K $\times 10^{12}$ dynes/cm² | $C_{11}$ = 1.84 $C_{12}$ = .662 $C_{44}$ = .220 |
| Linear Thermal Expansion 295 K | $22 \times 10^{-6}$/ K |
| Low Temperature Heat Capacity | |
| Optical Modes 295 K | $\omega_{TO}$ = 202 cm⁻¹, $\omega_{LO}$ = 384 cm⁻¹, |
| Reststrahlen | 49.5 $\mu$m |

Figure 4.12   Properties of single crystal cadmium fluoride (from Optovac, 1982).   Reprinted by permission of Optovac, Inc.

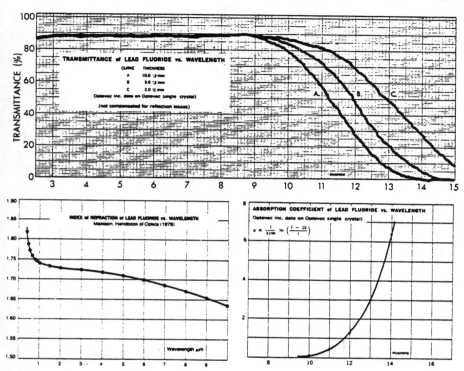

## General Physical Properties

| | |
|---|---|
| Structure | Cubic (fluorite); cleaves on 111 planes |
| Specific Gravity | 8.24 at 20°C : 7.74 at 23.4°C : 7.763 at 18°C |
| Melting Point | 822°C |
| Useful Transmission Range | 0.29 to 11.6 microns |
| Reflection Loss | 12% for two surfaces at 4 microns |
| Solubility | .064 gm per 100 gm water at 20°C |
| Hardness | 3. Moh approx. |
| Reciprocal Dispersive Power | 28.6 |
| Optical Modes 295 K | $\omega_{TO} = 102$ cm$^{-1}$. $\omega_{LO} = 337$ cm$^{-1}$ |
| Reststrahlen | 98 $\mu$m |

### Index of Refraction

| Wavelength $\mu$m | n | Wavelength $\mu$m | n |
|---|---|---|---|
| 0.30 | 1.93665 | 0.90 | 1.74455 |
| 0.40 | 1.81804 | 1.00 | 1.74150 |
| 0.50 | 1.78220 | 3.00 | 1.72363 |
| 0.60 | 1.76489 | 5.00 | 1.70805 |
| 0.70 | 1.75502 | 7.00 | 1.68544 |
| 0.80 | 1.74879 | 9.00 | 1.65504 |

Figure 4.13   Properties of single crystal lead fluoride (from Optovac, 1982). Reprinted by permission of Optovac, Inc.

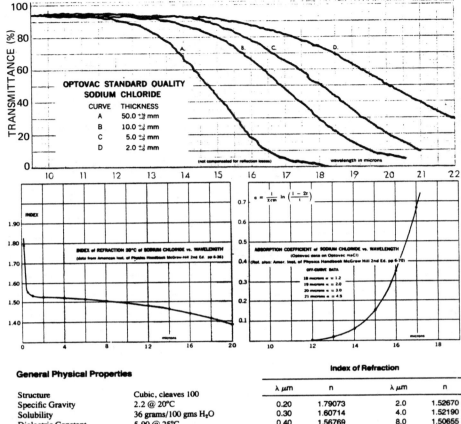

## General Physical Properties

| | |
|---|---|
| Structure | Cubic, cleaves 100 |
| Specific Gravity | 2.2 @ 20°C |
| Solubility | 36 grams/100 gms $H_2O$ |
| Dielectric Constant | 5.90 @ 25°C |
| Melting Point | 801°C |
| Thermal Conductivity | 0.0155 cal/sec cm°C @ 16°C |
| Thermal Expansion | $44 \times 10^{-6}/°C$ −50 to 200°C |
| Specific Heat | 0.204 cgs @ 0°C |
| Useful Transmission | 0.2 to 15$\mu$m |
| Hardness | 18.7 Knoop [100] <br> 16.0 Knoop [110] |
| Elastic Constants <br> $\times 10^{12}$ dynes/cm$^2$ | $C_{11}$ = .485 <br> $C_{12}$ = .123 <br> $C_{44}$ = .126 |
| Young's Modulus | $5.8 \times 10^6$ lb/in$^2$ |
| Rigidity Modulus | $1.83 = 10^6$ lb/in$^2$ |
| Bulk Modulus | $3.53 \times 10^6$ lb/in$^2$ |
| Poisson's Ratio | 0.252 |
| Optical Modes | $\omega_{TO}$ = 164 cm$^{-1}$ <br> $\omega_{LO}$ = 290 cm$^{-1}$ |
| Reststrahlen | 52 $\mu$m |

### Index of Refraction

| $\lambda\ \mu$m | n | $\lambda\ \mu$m | n |
|---|---|---|---|
| 0.20 | 1.79073 | 2.0 | 1.52670 |
| 0.30 | 1.60714 | 4.0 | 1.52190 |
| 0.40 | 1.56769 | 8.0 | 1.50655 |
| 0.50 | 1.55175 | 10.0 | 1.49482 |
| 0.70 | 1.53881 | 12.0 | 1.48004 |
| 0.90 | 1.53366 | 16.0 | 1.44001 |

Figure 4.14   Properties of single crystal sodium chloride (from Optovac, 1982). Reprinted by permission of Optovac, Inc.

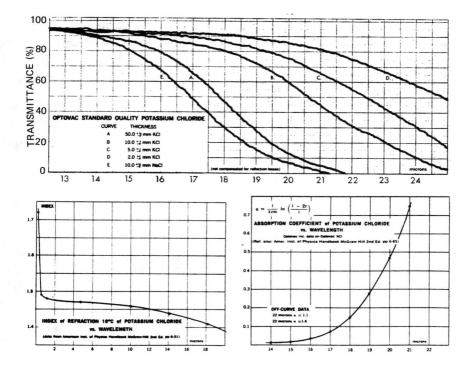

### Index of Refraction

### General Properties

| Structure | Cubic, cleaves on (100) planes |
| --- | --- |
| Specific Gravity | 1.9865 at 28°C |
| Reflection Loss | 6.8% for two surfaces at 10 microns |
| Dielectric Constant | 4.86 at 20°C |
| Melting Point | 776°C |
| Thermal Conductivity | $1.56 \times 10^{-2}$ cal/(cm sec C°) at 42°C |
| Thermal Expansion | $36 \times 10^{-6}$/C° in range 20°C to 60°C |
| Specific Heat | 0.162 at 0°C |
| Useful Transmission Range | .2 to 20 microns |
| Hardness | 7.2 Knoop [110] |
| | 9.3 Knoop [100] |
| Elastic Coefficients $\times 10^{12}$ dynes/cm$^2$ | $C_{11} = .398$ |
| | $C_{12} = .062$ |
| | $C_{44} = .063$ |
| Elastic Modulii | |
| Young's Modulus | $4.3 \times 10^6$ lb/in$^2$ |
| Rigidity Modulus | $0.906 \times 10^6$ lb/in$^2$ |
| Bulk Modulus | $2.52 \times 10^6$ lb/in$^2$ |
| Reststrahlen | 63 $\mu$m |
| Optical Modes | $\omega_{TO} = 142$ |
| | $\omega_{LO} = 214$ |

| Wavelength $\mu$m | n | Wavelength $\mu$m | n |
| --- | --- | --- | --- |
| 0.185409 | 1.82710 | 2.3573 | 1.474751 |
| 0.211078 | 1.67281 | 4.7146 | 1.471122 |
| 0.308227 | 1.54136 | 8.2505 | 1.462726 |
| 0.410185 | 1.50907 | 20.4 | 1.389 |
| 0.88308 | 1.481422 | 24.9 | 1.336 |

Figure 4.15 Properties of single crystal potassium chloride (from Optovac, 1982). Reprinted by permission of Optovac, Inc.

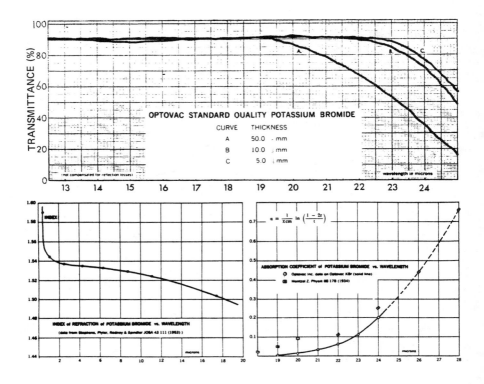

## General Physical Properties

| Structure | Cubic, cleaves on (100) planes |
|---|---|
| Specific Gravity | 2.75 at 23°C |
| Dielectric Constant | 4.90 in range 100 CPS to $1 \times 10^{10}$ CPS at 25°C |
| Melting Point | 730°C |
| Thermal Conductivity | $1.15 \times 10^{-2}$ cal/(cm sec C°) at 46°C |
| Thermal Expansion | $43 \times 10^{-6}$/C° in range from 20°C to 60°C |
| Specific Heat | 0.104 at 0°C |
| Useful Transmission Range | .23µ to 25µ |
| Reflection Loss | 8.4% for two surfaces at 10 microns |
| Solubility | 53.48 gm/100 gm water at 0°C |
| Hardness | 5.9 Knoop [110] 7.0 Knoop [100] |
| Elastic Coefficients $\times 10^{12}$ dynes/cm² | $C_{11} = .345$ $C_{12} = .054$ $C_{44} = .051$ |
| Elastic Modulii | |
| Young's Modulus | $3.9 \times 10^6$ lb/in² |
| Rigidity Modulus | $0.737 \times 10^6$ lb/in² |
| Bulk Modulus | $2.18 \times 10^6$ lb/in² |
| Reststrahlen Peak | 80 µm |

### Index of Refraction

| Wavelength µm | Index | Wavelength µm | Index |
|---|---|---|---|
| 0.404656 | 1.589752 | 8.662 | 1.52903 |
| 0.546074 | 1.563928 | 11.862 | 1.52200 |
| 0.706520 | 1.552447 | 14.98 | 1.51280 |
| 1.36728 | 1.54061 | 19.91 | 1.49288 |
| 2.440 | 1.53733 | 25.14 | 1.46324 |
| 4.258 | 1.53523 | | |

Figure 4.16   Properties of potassium bromide (from Optovac, 1982). Reprinted by permission of Optovac, Inc.

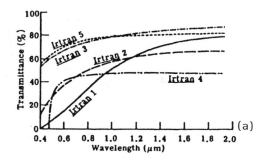

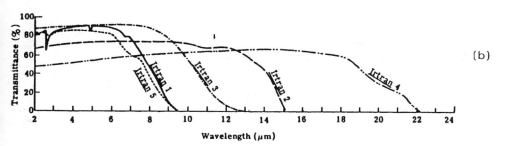

Figure 4.17    (a) Transmission of Irtran 1 through 5 (2.0 mm thickness).  (b) Transmission of Irtran materials (2.0 mm thickness). Irtran is a registered trademark of the Eastman Kodak Co. (from Wolfe and Zissis, 1978).   Reprinted with permission of W. L. Wolfe and G. J. Zissis, *The Infrared Handbook*, Environmental Research Institute of Michigan, 1978.

Properties of Bulk Inorganic Optical Materials

**Moduli of Rupture and Elasticity of
KODAK IRTRAN Materials
(in pounds per square inch)**

| KODAK Material | Modulus of Rupture (p s i) | Modulus of Elasticity (10⁶ p s i) | Temperature °C |
|---|---|---|---|
| IRTRAN 1 | 21,800<br>9,000—10,000 [a] | 16.6<br>16.6—13.5 [a] | 25<br>500 |
| IRTRAN 2 | 14,100<br>13,500<br>11,000 | 14.0<br>10.6<br>10.7 | 25<br>250<br>500 |
| IRTRAN 3 | 5,300<br>9,000 | 14.3<br>14.0 | 25<br>500 |
| IRTRAN 4 | 6,000<br>7,500 [b] | 10.3<br>— | 25<br>25 |
| IRTRAN 5 | 19,200<br>13,000<br>6,700 | 48.2<br>—<br>— | 25<br>500<br>900 |
| IRTRAN 6 | 4,600<br>4,500<br>5,900 | 5.2<br>5.3<br>4.5 | —196<br>25<br>100 |

[a] 5 to 25 minutes exposure
[b] Material processed for optimum grain structure

**Hardness Values of KODAK IRTRAN Materials**

| Material | Hardness Value — Knoop | Hardness Value — Approximate Moh | Common Comparison |
|---|---|---|---|
| IRTRAN 1 | 575 | 6 | About as hard as soft steel |
| IRTRAN 2 | 355 | 4.5 | A little harder than a penny |
| IRTRAN 3 | 200 | 4 | About as hard as a penny |
| IRTRAN 4 | 150 | 3 | About as hard as a marble |
| IRTRAN 5 | 640 | 6.5 | About as hard as a steel file |
| IRTRAN 6 | 45 | 2 | About as hard as hard coal |
| Diamond | | 10 | Hardest known material |
| Sapphire | 1370 | 9 | |
| Silicon | 1150 | 7 | |
| Germanium | 690 | 6.5 | |
| Fused silica | 480 | | |
| Arsenic Tri- sulphide | 110 | | |
| KRS-5 | 40 | | |
| Sodium Chloride | 17 | | |
| Optical glass | 400– 700 | | |
| Pyrex | 480 | | |

Figure 4.18   Physical properties of the Irtran materials (from Eastman Kodak Co., 1971).   Reprinted from Kodak Publication No. U-72, *Kodak Irtran Infrared Optical Materials*, Eastman Kodak Company, Rochester, NY, 1971.

## 4.7 CHEMICALLY VAPOR-DEPOSITED (CVD) ZnS AND ZnSe

CVD zinc sulfide and zinc selenide can be prepared in a highly pure stoichiometric flaw-free polycrystalline microstructure in large pieces, greater than 12 in. in diameter by 1 in. thick. ZnS and ZnSe prepared by CVD have properties which are superior to those of materials made by the hot pressing technique. Selected typical optical and physical properties of these two important materials are shown in Figs. 4.19 and 4.20. ZnSe has a longer cutoff than the ZnS, but ZnS is harder and more resistant to mechanical surface damage. ZnS is also less costly. The high refractive index of these solids is associated with high reflection losses at the surfaces, as is evident in the external transmission curves shown. However, the bulk absorptance of the ZnSe is exceptionally low, $4 \times 10^{-4}$ cm at 3.8 and 5.25 μm.

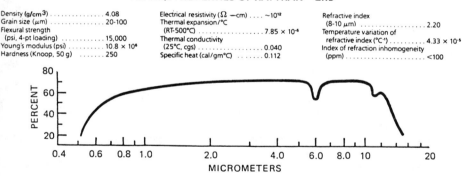

**TYPICAL PROPERTIES OF RAYTRAN™ ZnS**

| | | |
|---|---|---|
| Density (g/cm³) . . . . . . . . . . . . . . . 4.08 | Electrical resistivity (Ω —cm) . . . . ~10¹² | Refractive index |
| Grain size (μm) . . . . . . . . . . . . . . 20-100 | Thermal expansion/°C | (8-10 μm) . . . . . . . . . . . . . . . . . . 2.20 |
| Flexural strength | (RT-500°C) . . . . . . . . . . . . . . . . . 7.85 × 10⁻⁶ | Temperature variation of |
| (psi, 4-pt loading) . . . . . . . . . . . 15,000 | Thermal conductivity | refractive index (°C⁻¹) . . . . . . . . . . 4.33 × 10⁻⁵ |
| Young's modulus (psi) . . . . . . . . . 10.8 × 10⁶ | (25°C, cgs) . . . . . . . . . . . . . . . . . . 0.040 | Index of refraction inhomogeneity |
| Hardness (Knoop, 50 g) . . . . . . . 250 | Specific heat (cal/gm°C) . . . . . . . 0.112 | (ppm) . . . . . . . . . . . . . . . . . . . . . <100 |

Figure 4.19   Properties of ZnS (Raytran). Typical transmittance of Raytran ZnS (5 mm thick) (from Raytheon Co., 1980).

**TYPICAL PROPERTIES OF RAYTRAN™ ZnSe**

| | |
|---|---|
| Density (g/cm³) . . . . . . . . . . . . . . . 5.27 | |
| Grain size (μm)          . . . . . . . . . . . 70 | |
| Flexural strength (psi, | |
| 4-pt loading) . . . . . . . . . . . . . . . . 7500 | |
| Young's modulus (psi) . . . . . . . . . 9.75 × 10⁶ | |
| Poisson's ratio . . . . . . . . . . . . . . . 0.28 | |
| Hardness (Knoop, 50 gm) . . . . . . . 100 | |
| Resistivity ( Ω -cm) . . . . . . . . . ~ 10¹² | |
| Thermal expansion/°C, | |
| range 20-170°C . . . . . . . . . . . . . . 7.57 × 10⁻⁶ | |
| Thermal conductivity (25°C, cgs) . 0.043 | |

| | |
|---|---|
| Specific heat (cal/gm°C) . . . . . . . .0.081 | |
| Uncoated transmission 8-13 μm . >70 percent | |
| Transmission limits . . . . . . . . . . . 0.5 to 22 μm | |
| Refractive index | |
| 8-13 μm . . . . . . . . . . . . . . . . . . . 2.417 to | |
| 2.385 | |
| 10.6 μm . . . . . . . . . . . . . . . . . . 2.403 | |
| Index of refraction inhomogeneity | |
| (ppm, max) . . . . . . . . . . . . . . . . . <3 | |
| Temperature coefficient | |
| (dn/dT/°C) at 10.6 μm . . . . . . . . 6 × 10⁻⁵ | |

Extinction coefficient
@ 6328 Å (cm⁻¹) . . . . . . . . . . . . . <0.1
@ 1.06 μm (cm⁻¹) . . . . . . . . . . . <0.01
Bulk absorption coefficient
@ 2.77 μm (cm⁻¹) . . . . . . . . . . . 0.0007
@ 3.8 μm (cm⁻¹) . . . . . . . . . . . . 0.0004
@ 5.25 μm (cm⁻¹) . . . . . . . . . . . 0.0004
@ 10.6 μm (cm⁻¹) . . . . . . . . . . . 0.0005
Pulse damage threshold,
peak intensity @ 180 nsec
(GW/cm²)
@ 2.7 μm . . . . . . . . . . . . . . . . 0.59
@ 10.6 μm . . . . . . . . . . . . . . . . 0.64

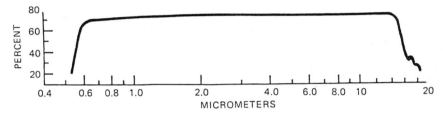

Figure 4.20    Properties of ZnSe (Raytran).   Typical transmittance of Raytran ZnSe (20 mm thick) (from Raytheon Co., 1981).

## 4.8   DIAMOND

Diamond is a unique substance and not conventionally thought of as an optical material.   However, commercially available optical quality natural diamond does exist.   Table 4.3 describes some of the properties of commercially available natural stones.

Diamonds can also be synthesized by the application of intense pressure and high temperature to carbonaceous precursors.   Many optical variations are possible by doping.   Figures 4.21 and 4.22 show spectra for a variety of natural and synthetic diamonds. Diamond resists corrosive agents and has a high thermal conductivity and a low coefficient of expansion, making it thermal shock resistant.

**Table 4.3  Properties of Commercial Diamond (Type IIa)[a]**

Optical Properties

Optical transmission: from 0.23 to beyond 200 μm. Note absorption bands between 2.5 and 6 μm. See transmission curve (Figure 4.22).

Index of refraction:   at 0.2265 μm   n = 2.7151
                       at 0.5461 μm   n = 2.4237
                       at 0.6563 μm   n = 2.4099

Absorption at 10.6 μm:     0.03% (±0.03%) per mm
                       or  0.003% (±0.003%) per 0.1 mm

Thermal Properties

Thermal conductivity:   26 W/°C cm at 20°C
                       120 W/°C cm at −190°C
   (compared with pure copper at 4 W/°C cm at 20°C),

Thermal expansion (linear):  $0.8 \times 10^{-6}$ per °C at 20°C
                             $0.4 \times 10^{-6}$ per °C at −100°C
                             $(1.5-4.8) \times 10^{-6}$ per °C from
                             100°−900°C

Specific heat (constant volume):   0.123 cal/g

Chemical Properties

Type IIa diamond is effectively free of nitrogen, which accounts for its enhanced thermal and optical properties over other types. Resistance to chemical attack: Diamond is extremely inert and is not affected by any acids or chemicals except high temperature oxidizing agents such as oxygen over 600°C and sodium nitrate over 450°C, high temperature carbide formers such as W, Ta, Ti, and Zr, and high temperature metallic carbon solvents such as Fe, Co, Mn, Ni, and Cr.

Mechanical Properties

Density:  $3.52 \, \text{g/cm}^3$ at 25°C

Hardness:  10 on Moh scale
           $5.7-10.4 \, \text{kg/mm}^2$ on Knoop scale, depending on direction and load (hardest material known)

Young's modulus:  approximately $10.5 \times 10^{12} \, \text{dyn/cm}^2$ in any direction

Bulk modulus:  $4.4-5.9 \times 10^{12} \, \text{dyn/cm}^2$ (highest of any material known)

Compressibility:  $1.7 \times 10^{-7} \, \text{cm}^2/\text{kg}$
                  (least compressible material known)

Tensile strength:  $(1.3-2.5) \times 10^{10} \, \text{dyn/cm}^2$

Shear strength:  $3 \times 10^9 \, \text{dyn/cm}^2$ torsion
                 $8.7 \times 10^{10} \, \text{dyn/cm}^2$ friction

[a]Source:  Oriel Corp. (1982).

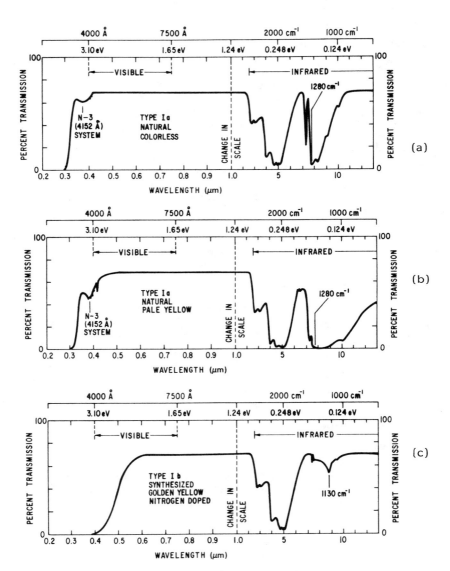

Figure 4.21   Spectra of diamonds of various types and colors—
ultraviolet to infrared: (a) type Ia natural—colorless; (b) type Ia
natural—pale yellow; (c) type Ib synthesized—golden yellow, nitrogen doped; (d) type Ib synthesized—greenish, nitrogen doped;
(e) type IIa synthesized—colorless; and (f) type IIb synthesized—
blue-boron-doped (from Chrenko and Strong, 1975).

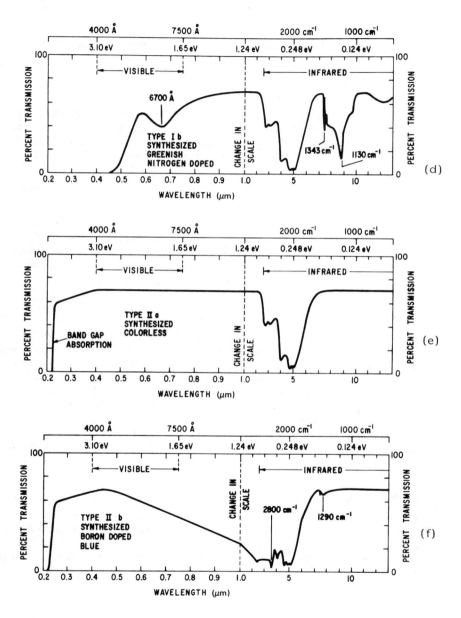

Figure 4.22 (continued from Fig. 4.21) Spectra of diamonds of various types and colors (from Chrenko and Strong, 1975).

## 4.9  SEMICONDUCTORS

Among the semiconductors of interest for infrared transmitting elements are

| | |
|---|---|
| Si | CdTe |
| Ge | GaAs |

The transmission of semiconductors drops off rapidly with temperature because of the increasing development of charge carriers due to thermal excitation. Typical properties and transmission data for the semiconductor materials listed above are given in Figs. 4.23–4.26.

### Silicon

Silicon transmits in the infrared up to 15 μm. It is hard and can be finished with ordinary glass-working equipment. It is in the cubic diamond structure and can be used in single or polycrystalline form with a minimum of scattering. However, due to its semiconducting nature absorption goes up rapidly with temperature.

### Germanium

Germanium is used in the 2- to 12-μm spectral region. Ge is non-hygroscopic, has good thermal conductivity, and excellent hardness and strength. Ge is a popular choice as a substrate for a variety of filters as well as for monolithic optical elements. Ge also has the cubic crystal structure. One disadvantage of Ge with respect to Si is its greater tendency to become opaque at elevated temperatures due to the relatively low band gap of Ge relative to Si (0.75 versus 1.14 eV at room temperature). This band gap difference results in easier stimulation of electrons from the valence band to the conduction band for Ge.

### Cadmium Telluride (CdTe)

Cadmium telluride has an IR bandpass from 1 to about 25 μm. It has a low thermal conductivity and is very soft. It is an excellent choice for a filter substrate in the 12- to 25-μm region, in which many other materials have decreased and variable transmittance due to the presence of absorption bands.

|                          | Si                          | Ge                          |
|--------------------------|-----------------------------|-----------------------------|
| Melting point            | 1420°C                      | 936°C                       |
| Density (at 25°C)        | 2.33 g/cm$^3$               | 5.323 g/cm$^3$              |
| Thermal expansion coefficient | 4.2 × 10$^{-6}$/°C     | 6.1 × 10$^{-6}$/°C          |
| Thermal conductivity     | 0.2 cal/sec cm°C            | 0.14 cal/sec cm°C           |
| Specific heat (0—100°C)  | 0.181 cal/g°C               | 0.074 cal/g°C               |
| Atomic weight            | 28.08                       | 72.60                       |
| Lattice constant         | 5.429 × 10$^{-8}$ cm        | 5.657 × 10$^{-8}$ cm        |
| Atoms per cm$^3$         | 4.99 × 10$^{22}$            | 4.41 × 10$^{22}$            |
| Volume compressibility   | 0.98 × 10$^{-12}$ cm$^2$/dyn | 1.3 × 10$^{-12}$ cm$^2$/dyn |
| Dielectric constant      | 12                          | 16                          |
| Energy gap               | 1.12 eV                     | 0.75 eV                     |
| Ionization energy        | about 0.04 eV               | about 0.01 eV               |
| Intrinsic resistivity (at 300 K) | 230,000 Ω cm        | 47 Ω cm                     |
| Electrons mobilities (at 300 K) | 1,500 cm$^2$/V sec    | 3,800 cm$^2$/V sec          |
| Holes mobilities (at 300 K) | 500 cm$^2$/V sec         | 1,800 cm$^2$/V sec          |

(a)

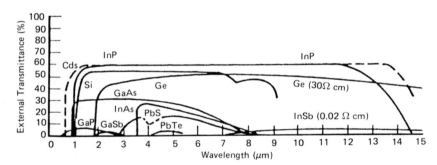

(b)

Figure 4.23 Some properties of silicon and germanium. (a) Physical properties of Si and Ge (from Alfa Crystals Datalog). (b) Transmission of cadmium sulfide, indium phosphide, silicon, germanium, gallium arsenide, gallium phosphide, gallium antimonide, indium arsenide, indium antimonide, lead telluride, and lead sulfide (from Wolfe and Zissis, 1978). Reprinted with permission of W. L. Wolfe and G. J. Zissis, *The Infrared Handbook*, Environmental Research Institute of Michigan, Ann Arbor, MI, 1978.

MATERIAL PROPERTIES

| | | | |
|---|---|---|---|
| Bulk Absorptivity (cm$^{-1}$ @ 10.6 $\mu$m) | $\leq$.030 @ 25C$^{\circ}$ | Temperature Change of Refractive Index (10$^{-6}$/C$^{\circ}$) | 277. |
| Thermal Conductivity (W/cm C$^{\circ}$) | .59 | Refractive Index @ 10.6 $\mu$m | 4.00 |
| Thermal Expansion Coefficient (10$^{-6}$/C$^{\circ}$) | 5.7 | Hardness (Knoop) | 692. |
| | | Rupture Modulus (psi) | 13500. |

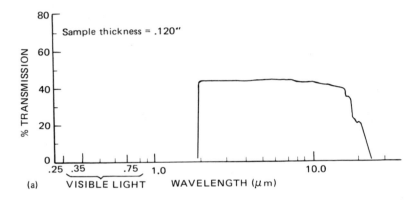

(a)

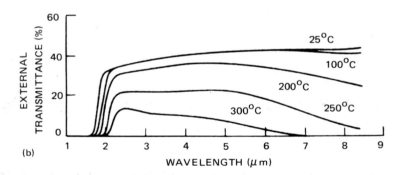

(b)

Figure 4.24   Properties of germanium.  (a) From II—IV, Inc.
(1983).  (b) Transmission of germanium for several temperatures
(1.15 mm thickness) (from Wolfe and Zissis, 1978).  Reprinted with
permission of W. L. Wolfe and G. J. Zissis, *The Infrared Handbook*,
Environmental Research Institute of Michigan, Ann Arbor, MI, 1978.

## MATERIAL PROPERTIES

| | | | | |
|---|---|---|---|---|
| Bulk Absorptivity (cm⁻¹ @ 10.6 μm) | ≤ .0018 | | Temperature Change of Refractive Index ($10^{-6}/C°$) | 107 |
| Thermal Conductivity (W/cm C°) | .06 | | Refractive Index @ 10.6 μm | 2.67 |
| Thermal Expansion Coefficient ($10^{-6}/C°$) | 5.9 | | Hardness (Knoop) | 45. |
| | | | Rupture Modulus (psi) | 3200. |

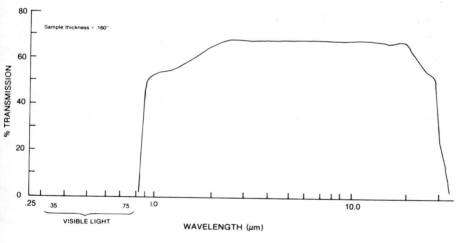

Figure 4.25   Properties of CdTe (from II–IV, Inc., 1983).

## Gallium Arsenide (GaAs)

Gallium arsenide is a relatively strong and hard optical material. GaAs optics are limited by crystal growth technology to dimensions in the 5- to 7-cm range. The material is nonhygroscopic and relatively stable against various solvents other than strong acids.

| | |
|---|---|
| Melting point °C | 1240 |
| Lattice constant: Å | 5.63 |
| Absorption edge: $\mu$m | 0.82 |
| Energy gap: eV | 1.35 |
| Density: g/cm$^3$ | 5.3 |
| Dielectric constant | 11.1 |
| Refractive index | 3.3 |
| Mobility @ 20°C: cm$^2$/V sec | 4000—5000 |
| Net carrier concentration/cm$^3$ | $> 75 \times 10^{16}$ |
| Resistivity: $\Omega$/cm at 20°C | 0.02—0.1 |
| Hall constant at 77 K: cm$^3$/V$\Omega$ | $> 100$ |
| Orientations (100) (110) (111) | |

Figure 4.26    Properties of GaAs (from II—IV, Inc., 1983).

## MATERIAL PROPERTIES

| | | | | |
|---|---|---|---|---|
| Bulk Absorptivity<br>(cm⁻¹ @ 10.6 μm) | ≤ .010 | | Temperature Change<br>of Refractive Index<br>($10^{-6}$/C°) | 149. |
| Thermal Conductivity<br>(W/cm C°) | .48 | | Refractive Index<br>@ 10.6 μm | 3.27 |
| Thermal Expansion<br>Coefficient<br>($10^{-6}$/C°) | 5.7 | | Hardness (Knoop) | 750. |
| | | | Rupture Modulus<br>(psi) | 20000. |

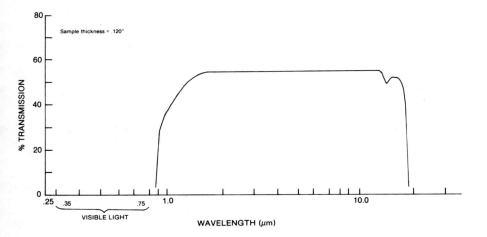

Figure 4.26    (Continued)

Table 4.4   Properties of Cz Sapphire[a]

---

**Physical**

| | |
|---|---|
| Density | $3.98$ g/cm$^3$ |
| Hardness | Mohs 9 (by definition) |
| | Knoop microindenter: |
| | 1600—2200 |
| Melting point | 2040°C |
| Compressive strength | 300,000 psi |
| Young's modulus | $(50-56) \times 10^6$ psi |
| Tensile strength | |
|   at 20°C | 58,000 psi (design minimum) |
|   at 500°C | 40,000 psi (design minimum) |
|   at 1000°C | 52,000 psi (design minimum) |
| Modulus of rupture | 65—100,000 psi |
|   (maximum bending stress | |
|   is orientation dependent) | |
| Modulus of rigidity | $27 \times 10^6$ psi |

**Thermal**

| | |
|---|---|
| Conductivity | |
|   (60° orientation) | |
|   at 0°C | 0.11 cal/cm°C sec |
|   at 100°C | 0.06 cal/cm°C sec |
|   at 400°C | 0.03 cal/cm°C sec |
| Specific heat | |
|   at 20°C | 0.10 cal/g |
| Heat capacity | |
|   at 20°C | 78 abs J/deg mol |
|   at 1000°C | 125 abs J/deg mol |
| Average coefficients of | |
|   thermal expansion | |
|   (60° orientation) | |
|   20—50°C | $5.8 \times 10^{-6}$ cm/cm °C |
|   20—500°C | $7.7 \times 10^{-6}$ cm/cm °C |

**Electrical**

| | |
|---|---|
| Volume resistivity | |
|   at 25°C | $10^{14}$ Ω/cm |

**Table 4.4** (Continued)

| Dielectric constant and dielectric loss tangent at 25°C | | | E⊥C | E∥C |
|---|---|---|---|---|
| K | | 1 MHz | 9.39 | 11.58 |
| | | 3 GHz | 9.39 | 11.58 |
| | | 8.5 GHz | 9.39 | 11.58 |
| tan δ | | 1 MHz | 0.0001 | < 0.0001 |
| | | 3 GHz | <0.0001 | < 0.0001 |
| | | 8.5 GHz | <0.00002 | < 0.00005 |

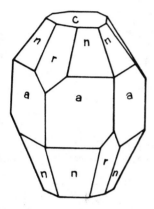

Sapphire, single crystal aluminum oxide, is classed as a rhombohedral structure but is normally indexed on hexagonal axes. Optically, it is a negative, uniaxial crystal in the visible; it exhibits anisotropy in its physical, thermal, and dielectric properties. An amphoteric semiconductor, sapphire's energy band gap, approximately 10 eV, which is one of the largest for oxide crystals, permits useful optical transmission to extend from about 1450 Å to 5.5 μm.

[a]Source: Union Carbide Electronics Div. (1980). Reprinted by permission of Union Carbide Electronic Materials.

## 4.10 SAPPHIRE

Sapphire is the single crystal form of aluminum oxide, $Al_2O_3$. Its energy band gap is about 10 eV, permitting useful transmission from approximately 145 nm to 5.5 μm. Some physical properties of single crystal Czochralski grown sapphire are displayed in Table 4.4. Figure 4.27 shows the index of refraction of sapphire. Sapphire has a birefringence of 0.0008 in the visible.

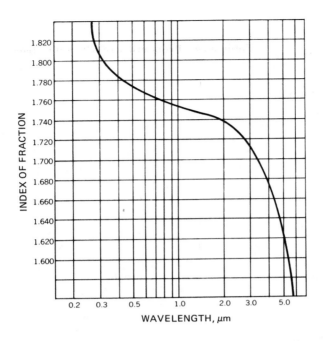

**Figure 4.27** Index of refraction of sapphire (from Union Carbide, 1980). Reprinted by permission of Union Carbide Electronic Materials.

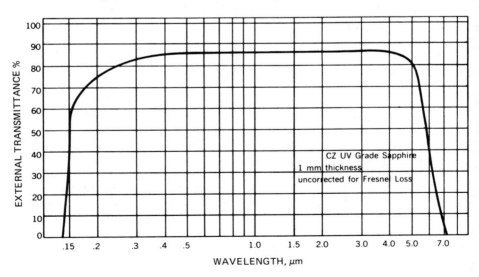

**Figure 4.28** Transmission of sapphire (1 mm thick) (from Union Carbide, 1980). Reprinted by permission of Union Carbide Electronic Materials.

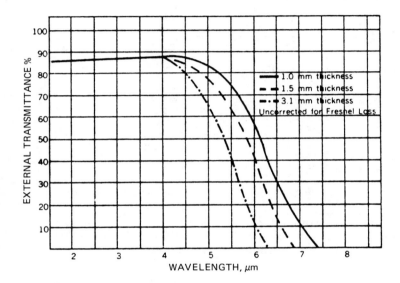

**Figure 4.29** Transmission of sapphire (various thicknesses) (from Union Carbide, 1980). Reprinted by permission of Union Carbide Electronic Materials.

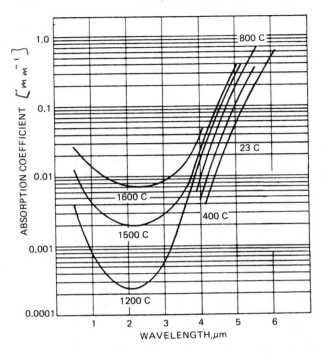

Figure 4.30    Absorption coefficient of sapphire in the infrared as a function of temperature (from Union Carbide, 1980). Reprinted by permission of Union Carbide Electronic Materials.

The thermal coefficient of the index of sapphire, dn/dT, is approximately $13 \times 10^{-6}$°C in the visible. The transmission of sapphire for a 1-mm-thick plate is indicated in Fig. 4.28 and for other thicknesses in Fig. 4.29. The effect of temperature on the absorption coefficient of sapphire is shown in Fig. 4.30.

## REFERENCES

Alfa Crystals, Ventron Materials Division, *Alfa Crystals Datalog*, Bradford, PA.

R. M. Chrenko and H. M. Strong, *Physical Properties of Diamond*, Report No. 75CRD089, General Electric Company, Schenectady, NY, 1975.

Eastman Kodak Company, *Kodak IRTRAN Infrared Optical Materials*, Kodak Publication U-72, Rochester, NY, 1971.

N. C. Fernelius, G. A. Graves, and W. L. Knecht, Characterization of candidate laser window materials, *Proc. Society of Photo-Optical and Instrumentation Engineers 297*, 1981.

D. E. Gray, *American Institute of Physics Handbook*, McGraw—Hill Book Company, Inc., New York, 1963, Sec. 6.

The Harshaw Chemical Company, *Crystal Optics*, Catalog OP 101, Solon, OH, 1982.

Ohara Optical Glass, Inc., *General Catalogue*, Wachung, NJ, Sasamihara, Japan, 1982.

Optovac, Inc., *Optical Crystals by Optovac*, Handbook 82, North Brookfield, MA, 1982.

Oriel Corp., *Oriel Corporation Complete Catalogue of Optical Systems and Components*, Stamford, CN, 1982.

Raytheon Co., *Raytheon Infrared Materials*, Waltham, MA, 1980.

Schott Glass Technologies, Inc., *Optical Glass*, Duryea, PA, 1982.

II-VI Inc., *II-VI Infrared Optics*, Saxonburg, PA, 1983.

Union Carbide Electronics Div., *Properties and Applications of Cz Sapphire*, San Diego, CA, 1980.

W. L. Wolfe and G. J. Zissis, *The Infrared Handbook*, Environmental Research Institute of Michigan, Ann Arbor, MI, 1978.

# 5
# PROCESSING OF OPTICAL MATERIALS

## 5.1 INTRODUCTION

This chapter briefly describes representative methods for making
bulk glass, ceramic crystalline materials, and halides. Powder
processes, single crystal techniques, chemical vapor deposition
(CVD), and molten salt bath approaches will be reviewed. The
fabrication and application of plastic optics will be dealt with in
Chap. 6.

The reason for including a discussion of processes is to provide
the user of optical materials an appreciation of the many variable
and controllable parameters associated with the manufacture of high
quality materials for the optic arts.

## 5.2 GLASS PROCESSES

The manufacture of glass is an ancient art. Like many ancient
arts, its development has been continuous throughout man's history up to the present day. The possible varieties of glass are
virtually infinite and, indeed, manufacturers of optical glass
offer hundreds of compositions, heat treatments, and surface
processes.

There are many manufacturers of high quality optical glass including Corning Glass Works, Hoya Optics, Inc., Schott Glass Technologies, Inc., and Ohara Optical Glass, Inc. The manufacture of
glass can be divided into three general areas: batching, melting,
and forming. Batching is formulation, melting is preparation into
a usable homogeneous material, and forming is the creation of a
specific shape. Batching includes the selection, formulation and
mixing of the raw ingredients and their delivery to the melter.
Melting includes the choice of refractory high-temperature liner
material in the furnace, the design of the furnace, and its operation during the melting and refining of the glass prior to forming.

Forming includes the delivery of the glass to the exit orifice of
the furnace and the process employed to shape the glass.

Glass has no crystalline structure. In fact, one of the key problems of the glassmaker is to avoid devitrification (crystallization)
of the glass during any part of the process. This implies that the
compositions and temperatures must be maintained in regions where
the glass has minimal tendency to nucleate and grow into a crystalline form. In addition, since the forming operations are extremely
sensitive to changes in viscosity, the glass temperatures must be
carefully controlled during that operation, usually within $\pm 10°F$.

## Batching

Raw materials occur as natural minerals processed to glass industry
specifications or are synthesized. For example, sand is a naturally
occurring crystalline quartz, while soda ash (NaO) is synthesized
from limestone and salt. Raw materials supplied to the batch house
must be selected for their chemical composition, uniformity of that
composition, particle size and particle size distribution, and free-
dom from contaminants.

A particularly troublesome impurity in the sand is $Fe_2O_3$. Not
only is this oxide of iron a strong absorber of light in the finished
part, but it inhibits heat transfer in the glass melt itself. Extremely pure quartz sands must be used for optical glasses.

If the particle size of the sand varies from batch to batch,
then the charge melts more slowly or more quickly depending on
particle size. This upsets the heat balance of the furnace and
leads to nonuniformity of the final product and also, in the case of
coarse particles, can lead to inclusions of unmelted quartz in the
final product.

## Melting

Batch melting and glass fining typically occur in one furnace with
three sections: the batch inlet section, the melter, and the forehearth, as indicated in Fig. 5.1. The choice of refractories which
line the furnace must withstand the severe chemical attack of the
melted glass as well as erosion by glass movement. The most severe wear usually occurs at the liquid—gas interface and, after a
time, cuts completely through the refractory walls. The refractories must be able to withstand the direct flame impingement and
combustion products of the fuels employed.

Another important consideration in the choice of the furnace
refractories is the requirement to minimize dissolution of substances

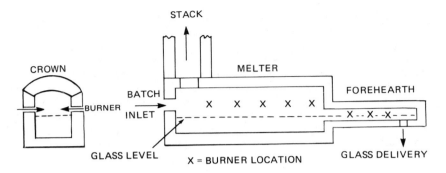

**Figure 5.1**   Glass furnace schematic.

into the glass which may degrade the desired optical properties of the end products.

The major constituents of furnace refractories are compositions of $SiO_2/Al_2O_3$, $ZrO_2/SiO_2$, or $SiO_2$ (silica). Silica is generally used for the crown of the furnace. The crown is the roof of the melting chamber and is constructed as a self-supporting arch. These refractories must be of high purity to avoid unacceptable contamination of the carefully compounded glass.

The delivery of the glass to the forming operation is through the forehearth. The temperature in the forehearth is lower than that of the melter. The small volume of the forehearth permits easier temperature adjustments.

### Forming

The four main processes for making glass objects in the optical glass industry are (1) direct process, (2) strip process, (3) casting process, and (4) clay pot process. In each process the glass batch or glass feed stock is melted in a gas fired or electric furnace. These processes are schematically shown in Figs. 5.2−5.5.

In the direct process, a forming machine is fed directly from the forehearth. In one such arrangement, the glass stream leaving the forehearth is cut into appropriately sized pieces, called gobs, after the molten stream has cooled to a plastic state. A high speed pressing machine then converts the glass gob into the desired blank. The blank is then carried through an annealing furnace, called a lehr. After the annealing step, the blank is ready for the finishing operations. This is the process used for high production parts such as ophthalmic blanks and camera lenses.

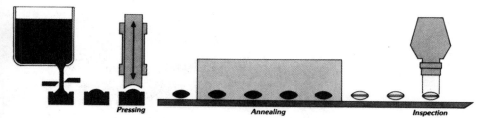

**Figure 5.2** Direct press process (from Ohara, 1982). Reprinted by permission of Ohara Optical Glass, Inc.

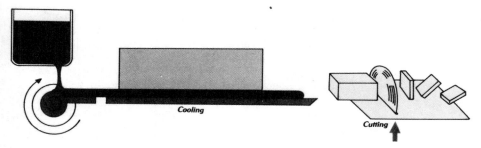

**Figure 5.3** Strip process (from Ohara, 1982). Reprinted by permission of Ohara Optical Glass, Inc.

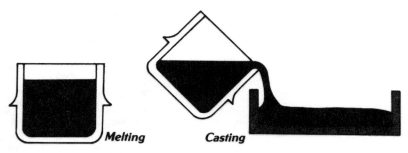

**Figure 5.4** Casting process (from Ohara, 1982). Reprinted by permission of Ohara Optical Glass, Inc.

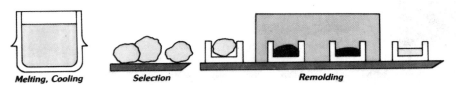

**Melting, Cooling**          **Selection**                    **Remolding**

Figure 5.5   Clay pot process (from Ohara, 1982). Reprinted by permission of Ohara Optical Glass, Inc.

The strip process continuously forms a strip from the molten glass in a roller device. The strip is continuously fed into a lehr for annealing. After cooling, the strip is cut into the desired length and is ready for the finishing operations. This is also a highly cost effective process.

The casting process uses a mold into which the glass is cast. The billet is cut into the proper shapes for the subsequent operations. Annealing takes place at some convenient point in the finishing process.

The clay pot process is relatively primitive. Glass is remelted in a low cost, easily broken, clay pot and allowed to cool. Then the pot is broken and good chunks of glass are selected for further operations. Each piece is put into a refractory mold and reheated and reformed. It is then ready for further operations.

Materials made by the strip, casting and clay pot processes are cut to shape and annealed, or pressed to shape and annealed to provide optical blanks for final grinding and polishing operations.

The manufacturing processes used in the glass industry are described in detail by Tooley (1953).

## 5.3 FUSED QUARTZ

Fused quartz is made by melting very high purity crystalline quartz such as Brazilian quartz which is mined. Quartz from surface sands is rarely as pure as the mined Brazilian quarts, which has metal impurity levels in the few tens parts-per-million range. However, for very exotic applications even this is not good enough, and synthetic amorphous silica glass is made by a number of chemical vapor deposition (CVD) processes. The CVD process starts with purified silicon tetrachloride and yields an amorphous silica glass with about 1 ppm of metallic impurities. This material can be fused into a product known as fused silica. Thus fused quartz and fused silica are distinctly different products.

Another significant contaminant is the OH⁻ ion, which produces a strong absorption band at 2720 nm and weak bands at 1380 and 2200 nm. By means of electrical fusion under vacuum or flame fusion in a water vapor-free flame, the OH⁻ ion can be suppressed to less than 5 ppm. The major suppliers of synthetic fused silica in the United States are Amersil Inc., Dynasil Inc. and the Corning Glass Works. The General Electric Co. produces fused quartz by electrical fusion of natural quartz. These materials were discussed earlier in Sec. 4.5.

Typical transmittance data for a number of varieties of fused silica were displayed in Fig. 2.11.

## 5.4 CERAMIC POLYCRYSTALLINE MATERIALS— POWDER METHOD

Starting materials for ceramic polycrystalline bodies can be minerals or polymeric precursors which are converted to ceramic powders by thermal processes. Traditional ceramic processing starts with mined minerals, much like in glass manufacture. However, the stringent optical requirements of modern optics permit very little leeway in absorption or scattering, and the preparation of highly purified starting powders is essential for high optical quality materials. The starting materials must be chemically pure as well as phase pure. Second phase inclusions generally differ in index of refraction from the host material thereby inducing scattering at each interface. The processes employed to achieve the optical blank, in general, must not induce the formation of multiphases. A flow diagram for fabrication of ceramic bodies from powder starting materials is shown in Fig. 5.6.

Small average particle size is desired in most cases, which helps to achieve a final small grain size. This enhances strength and influences scattering in birefringent materials such as $MgF_2$. The particle size distribution must be controlled to facilitate the elimination of porosity in the final blank. Small particle size enhances reactivity and encourages densification of the powder during thermal processing.

In the case of transparent alumina for sodium vapor lamp envelopes, the average starting particle size is 0.3 μm although the

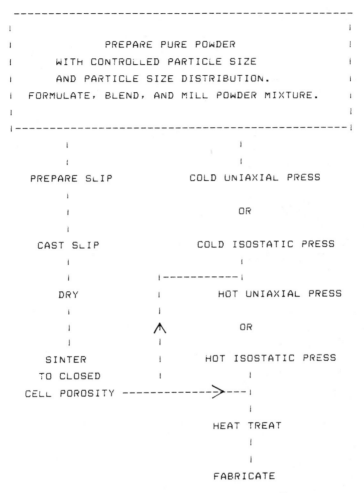

**Figure 5.6** Flow chart—powder process for ceramic.

final grains after sintering are of the order of 10 μm and near theoretical density is achieved.

An example of a high purity polymeric precursor approach is in the use of alkoxides to prepare mullite. In the alkoxide process metal alkoxides such as $Si(OR)_4$ and $Al(OR)_3$ are hydrolyzed (reacted with $H_2O$) in a controlled manner to form active polymerizable species in soluble form. R is the symbol for an alkoxide group, $C_nH_{2n + 1}$. The suitably polymerized compound is dissolved in water or alcohol and subsequently dried to produce a very fine amorphous powder with the desired chemistry. The chemical reactions are summarized in the following (Prochazka and Klug, 1984):

(a) $6Al(OR')_3 + 2\ Si(OR'')_4 + xH_2O =$
$$Al_6Si_2O_{13 - x/2}\ (OR)x + xRO_4$$

(b) $Al_6Si_2O_{13 - x/2} = Al_6Si_2O_{13} + (x/2)H_2O + alkene$

where R, R', and R" stand for alkyls. In this case, the desired powder is mullite, $Al_6Si_2O_{13}$.

During the ensuing thermal processing the crystallization of the amorphous powder can be controlled to result in a very fine-grained solid body exceptionally free of impurities.


## Consolidation of Powders

The major thermal processes for converting powder to a polycrystalline solid are sintering (S), hot pressing (HP), and hot isostatic pressing (HIP). In each case thermal energy is used to bond the individual particles of powder together under varying pressure conditions. In sintering, no external force is applied. In HP uniaxial external force is applied to the powder contained in a confining mold, whereas in the HIP process the body is subjected to isostatic pressure during the thermal process. The processes are quite general, but for optical quality material the process control must be much tighter than for structural ceramics.

The starting powders are mixed and milled to achieve a highly homogeneous mixture of fine particles with a precisely controlled distribution of particle sizes. In this process care must be taken to avoid the introduction of impurities from the milling media.

To prepare for sintering, the powder is mixed with a binder which can be organic or inorganic and then cold pressed either uniaxially or isostatically to achieve a high density "green" body. The green body must have adequate strength to withstand the various handling stresses of the consolidation process.

Another way to prepare the powder for sintering is to employ slip casting. Typically the slip is a suspension of the powder in water. The slip is cast into a mold made of a water absorbing material such as porous plaster. Figure 5.7 illustrates the simple drain casting technique.

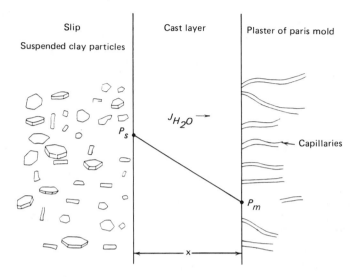

Figure 5.7 Slip casting. Schematic representation of the formation of a slip-cast layer formed by the extraction of water by capillary action of a plaster of paris mold (from Kingery, Bowen, and Uhlmann, 1976). Reprinted with permission from W. D. Kingery, H. K. Bowen, and D. R. Uhlmann, *Introduction to Ceramics*, John Wiley and Sons, Inc., 1976.

As mentioned above, densification or consolidation is effected by sintering, hot pressing or by a combination of sintering followed by hot isostatic pressing.

Sintering is a thermal process which is conducted at elevated temperature in air or in a controlled environment depending on the chemistry of the charge. It is the most economical process for the achievement of ceramic bodies.

In the hot pressing process, the powder without binder is confined in a die and subjected to high pressure and temperature on a prescribed schedule. The pressure is uniaxial and can lead to property variation in the longitudinal and the transverse directions. Often, contamination from the die must be removed by mechanical removal of the contaminated material or by thermal treatment such as oxidizing of carbon contaminants.

There are two approaches to hot isostatic pressing (HIP) of optical materials. In the first, the body is partially sintered until the porosity has been reduced to a closed (noninterconnecting) pore morphology. Then the partially sintered body is subjected to HIP. Under these circumstances the body, at temperature, is consolidated by the isostatically applied pressure of many thousands of pounds per square inch.

In the more direct but more difficult approach, the powder is placed in a sealed container, which becomes plastic without reptur-ing at the processing temperature, and then the sealed ampule is subjected to the hot isostatic pressure process. After the body is densified, various heat treatments are employed to control grain size, chemical oxidation-reduction reactions, and residual strains.

### Critical Problems

When synthesizing polycrystalline ceramic bodies, it is essential that the following objectives be attained if an optically useful mate-rial is to be achieved:

Eliminate porosity or reduce size of pores to a small fraction of the wavelength of interest.

Minimize impurities which may contaminate the material at any stage of the process.

Minimize the grain size of all birefringent and intentionally multi-phase materials. Microstructural features should be a small fraction of the wavelength of interest.

Eliminate unintentional second phases.

Minimize concentration of second phases intentionally incorporated and distribute them uniformly.

### 5.5 CHEMICAL VAPOR DEPOSITION (CVD)

The CVD processes for amorphous silica and for optical fibers have been discussed briefly earlier. The CVD process is also used with great success in making high optical quality polycrystalline zinc sulfide and zinc selenide. These are very useful IR transmitting materials. The basic chemical reactions for the process are

$$Zn_{vap} + H_2S_g = ZnS_{solid} + H_{2g}$$

$$Zn_{vap} + H_2Se_g = ZnSe_{solid} + H_{2g}$$

which occur at 600 to 800°C and pressures below 100 Torr. The process is indicated schematically in Fig. 5.8.

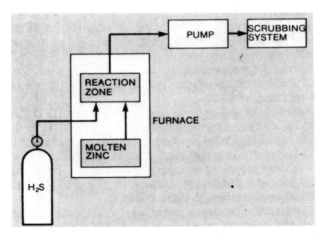

**Figure 5.8** Chemical vapor deposition (CVD) technique (from Taylor and Donadio, 1981). Chemical vapor deposition process for fabrication of ZnS; the production of ZnSe is similar, but uses $H_2Se$ rather than $H_2S$.

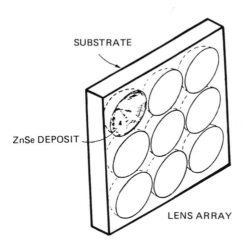

**Figure 5.9** Mandrel for producing an array of lenses (from Donadio, Connolly, and Taylor, 1981). R. N. Donadio, J. F. Connolly, and R. L. Taylor, New advances in chemical vapor deposited (CVD) infrared transmitting materials, *Emerging Optical Materials*, Proc. SPIE *297*, 1981, pp. 65–69.

The zinc vapor is obtained from a reservoir of liquid zinc, and the $H_2S$ and $H_2Se$ are introduced from storage bottles containing semiconductor grade purity gases. Chambers having as large as a 48 in. diameter are in use which make possible plates as large as 40 × 30 × 1 in. thick. A mandrel for producing an array of lenses is shown in Fig. 5.9. Very precise control of all process parameters is essential to achieve high optical quality material on a consistent basis.

Post deposition, high temperature and pressure treatments of ZnS can be used to improve the optical properties. The improvement in the visible part of the spectrum is believed to be due to porosity removal and the zinc leaching atmosphere of the heat treatment environment. Excess Zn is thought to enter the lattice during the CVD process.

## 5.6   SINGLE CRYSTALS

Single crystals have no grain boundaries to scatter light and, if birefringent, can be oriented so that the optic axis is in a favorable orientation. In addition to naturally occurring single crystals, many ways have been devised to synthesize them. Among the materials synthesized are sapphire ($Al_2O_3$), Si, Ge, $CaF_2$, NaCl, $MgF_2$, quartz ($SiO_2$), KCl, and GaAs.

### Single Crystals Grown from the Melt

Methods of growing single crystals from the melt are based on the principle that extraction of latent heat must be achieved while preventing the melt from supercooling to such an extent that competing crystals are nucleated. This requires extraction of heat through the crystal being grown. A seed crystal is used, one end in contact with a heat sink and the other in contact with the melt.

### Bridgman Method

There are a number of methods of growing single crystals from the melt. In the Bridgman method (Fig. 5.10), a mold with its long axis vertical is lowered through a furnace at a controlled rate so that solidification begins at the point of the mold where an initial seed crystal has been placed.

All of the conventional optical materials for applications in the ultraviolet and the infrared regions can be grown by the Bridgman technique. These include CsI, KBr, KI, NaB, NaF, $CaF_2$, LiF, Ge, and Si.

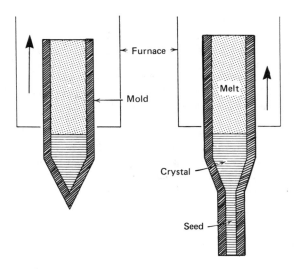

Figure 5.10    Bridgman method for growing single crystals (from Chalmers, 1964).    Reprinted with permission of B. Chalmers, *Principles of Solidification*, John Wiley and Sons, Inc., New York, 1964.


A horizontal boat can also be used as shown in Fig. 5.11.    The seed crystal is again placed at the narrow end of the boat, and solidification of a single crystal occurs as the mold moves from right to left through the furnace.

### Czochralski Method

The Czochralski method for growing single crystals is one of the best-known techniques.    The method is illustrated in Fig. 5.12.

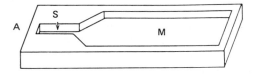

Figure 5.11    Crystal growth in a horizontal boat (from Chalmers, 1964).    Reprinted with permission of B. Chalmers, *Principles of Solidification*, John Wiley and Sons, Inc., New York 1964.

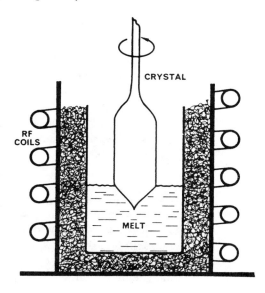

Figure 5.12   Czochralski method for growing single crystals (from Union Carbide, 1980).   Reprinted by permission of Union Carbide Electronic Materials.

A small seed crystal is brought in contact with the melt of the compound, slowly rotated as it is slowly withdrawn from the melt. As the seed is withdrawn, epitaxial growth of the crystal occurs.

The rate of upward motion is equal to the rate of crystal linear extension.   The rate of crystal growth is determined by the rate of latent heat extraction through the seed crystal to a heat sink. The purpose of the rotation is to maintain symmetry of the crystal. If there were no rotation, the shape would be unstable.

The crystallographic orientation of the crystal so grown can be controlled by the orientation of the seed crystal.   Sapphire boules having diameters up to 4 in. are made in this manner and are commercially available.

### Floating Zone Technique

A fourth method for growing single crystals from the melt is the floating zone technique (Fig. 5.13).   In this process, a solid polycrystalline rod of the substance is held vertically and a narrow zone is heated by induction or electron beam.   A narrow band or

**Table 5.1**  Guide to Standard Techniques Crystal Growth[a]

| Technique | Typical materials |
| --- | --- |
| Electron bombardment | Niobium, tantalum, tungsten, vanadium, molybdenum |
| Induction heating float zoning | Rare earths, chromium, nickel |
| Czochralski | Antimony, barium, tellurium, oxide crystals, alkali materials, silicon, germanium arsenide, gallium phosphide |
| Strain anneal | Iron, titanium, zirconium, thallium, cobalt/iron |
| Solid state electrolysis | Rare earths, titanium |
| Bridgman | Melting points to 1100°C: aluminum, copper, cadmium, gallium, lead, tin, zinc |
| (1) Graphite crucibles Bridgman (2) soft molds (MgO, $Al_2O_3$ or C powder) | Bismuth, antimony |
| Bridgman (3) refractory crucibles (MgO or $Al_2O_3$) | M.P. 1100 to 1800°C: nickel, cobalt, palladium, platinum, iron-silicon, nickel/iron |

| Typical sizes | As-grown external surface |
|---|---|
| To 12 mm diameter; to 250 mm long | Surfaces somewhat irregular, not necessarily perfectly circular or straight; no pores or inclusions |
| To 6 mm diameter | Surfaces somewhat irregular, not necessarily perfectly circular or straight; no pores or inclusions |
| Varies from material to material | No generalization possible, materials specified individually |
| Only supplied in stock sizes or machined from stock crystals | Smoothness of surface not changed by solid state techniques |
| 6 mm diameter × 50 mm long. Diameter normally to 62 mm (in some cases to 120 mm) | Cylindrical surface substantially smooth, straight and of regular cross section. Cylindrical surfaces reasonably smooth, corresponding to a carefully machined graphite mould |
| Length to 150 mm; also as slabs, bars, or spheres | Cylinders truly circular in section, but surfaces may contain very small pores, protrusions or graphite inclusions |
| Typically to 62 mm (to 115 mm for bismuth) | Cylinders not necessarily perfectly circular in section; somewhat rough surfaces possibly containing small powder inclusions |
| Melting point below 1750°C— up to 50 mm diam; melting point above 1750°C—up to 12 mm diam; length up to 125 mm | Surfaces generally smooth, but not perfectly circular in section or straight; may contain a few pores or inclusions of crucible material in surface |

**Table 5.1**  (Continued)

| Technique | Typical materials |
| --- | --- |
| Bridgman/Stockbarger | Lithium floride, sodium chloride, inorganic and optical grade crystals |
| Arc transfer technique | Nickel oxide, cobalt oxide, iron oxide |
| Verneuil | Oxide crystals |
| Flux melt | Oxides and complex oxides, e.g., spinels, garnets, silicates |

[a]Source:  Alfa Crystals/Ventron Materials Division, Bradford, PA.

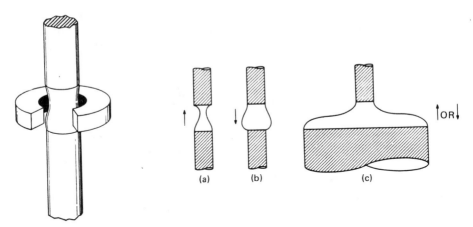

**Figure 5.13**   Floating zone technique for growing single crystals (from Pfann, 1966).  Reprinted with permission from W. G. Pfann, *Zone Melting*, John Wiley and Sons, Inc., New York, 1958. (a) Floating-zone technique (schematic).  Ring-shaped heater or induction coil produces molten zone which is held in place by sur- face tension.  (b) Stable shapes of molten zones; (*a*) upward- moving, (*b*) downward moving, (c) upward or downward moving.

| Typical sizes | As-grown external surfaces |
|---|---|
| Very large crystals can be produced | Surfaces generally smooth |
| 10 × 10 × 10 mm | As grown boule of irregular shape—may contain small grains at periphery |
| Varies with material | Nonporous, symmetrical |
| Generally small crystals up to 10 × 10 × 10 mm | As grown crystals of irregular shape |

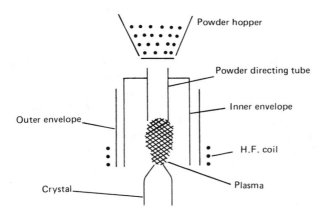

**Figure 5.14**  Verneuil flame fusion method for growing single crystals (from Pamplin, 1975). Verneuil apparatus using an open h.f. plasma torch. Reprinted with permission from B. R. Pamplin, *Crystal Growth*, Pergammon Press, Ltd., London, 1975.

zone of material melts but is held in place by surface tension forces. Starting at one end near the seed crystal, a new single crystal can be grown by moving the hot zone from end to the other.

A decided advantage of this method is that no container is required and crucible contamination is avoided. Single crystals of Ge and Si are made this way in large quantities for use in the semiconductor industry.

### Verneuil Vapor Process

Another group of processes for the production of single crystals is based on the condensation of vapors. One such technique is the Verneuil flame fusion process, illustrated in Fig. 5.14. Here a flame gives off combustion products of the correct composition, and these gases condense on a seed crystal thus initiating an epitaxial growth process, which leads to a single crystal boule.

Table 5.1 summarizes the variety of processes employed to grow single crystals. Pamplin (1975) has summarized crystal growing processes in a comprehensive treatise. Gilman (1963) has written a book covering the technology.

### REFERENCES

Alpha Crystals, Ventron Materials Div., *Alpha Crystal Datalog*, Bradford, PA.

B. Chalmers, *Principles of Solidification*, John Wiley and Sons, Inc., New York, 1964.

R. N. Donadio, J. F. Connolly, and R. L. Taylor, New advances in chemical vapor deposited (CVD) infrared transmitting materials, Proc. SPIE *297*, Bellingham, WA, 1981.

J. J. Gilman, *The Art and Science of Growing Crystals*, John Wiley and Sons, Inc., New York, 1963.

W. D. Kingery, H. K. Bowen, and D. R. Uhlmann, *Introduction to Ceramics*, John Wiley and Sons, Inc., New York, 1976.

Ohara Optical Glass, Inc., *Ohara Optical Glass*, Watchung, NJ, 1982.

B. R. Pamplin, *Crystal Growth*, Pergammon Press, New York, 1975.

W. G. Pfann, *Zone Melting*, John Wiley and Sons, Inc., New York, 1966.

S. Prochazka and F. J. Klug, US Patent No. 4,427,785, Optically translucent ceramic, 1984.

R. L. Taylor and R. N. Donadio, Vapor deposited IR materials, Laser Focus *17*(7), 1981.

F. V. Tooley, *Handbook of Glass Manufacture*, Ogden Publishing Co., New York, 1953, Vols. 1 and 2.

Union Carbide Electronic Materials, *Optical Properties and Applications of Cz Sapphire*, San Diego, CA, 1980.

# 6
# POLYMERS FOR OPTICS

## 6.1 INTRODUCTION

In high production, polymeric plastic materials are significantly less expensive than equivalent glass elements and they are less fragile. The cost of plastic elements in production ranges from a third to a twentieth of equivalent glass. In addition, the ability to mold the material to finished dimension, surface perfection, and figure makes aspheric elements and elements with integrally molded attachment fittings practical.

## 6.2 CHARACTERISTICS OF OPTICAL PLASTICS

Plastic elements weigh less than equivalent glass lenses, and more complex shapes can be designed and effectively manufactured. These driving forces have led to a significant plastic lens design and manufacturing technology.

In comparison to glass there are only a few plastics suitable for high quality optical application, and these are found in a narrow region of the index of refraction—Abbe number map. The basic materials are shown in Table 6.1.

Since prototype costs are prohibitive in plastic, it is common to use glass with optical properties close to those of the plastic for prototype development. Such glasses are K10 for acrylics, $TiF_5$ or $TiF_6$ (Schott or equivalent) for styrene-acrylonitrile (SAN), copolymer styrene-acrylic (NAS), polycarbonate, and styrene.

The negative aspects of plastic optical components can be tabulated under the following headings:

Table 6.1  Characteristics of Optical Plastics

| Relative cost | n | $\nu_d$ | Common name/trade name | %T[a] | °F[b] | Remarks |
|---|---|---|---|---|---|---|
| 4 | | | Allyl diglycol carbonate CR39 (PPG) ADC | | 212 <br> 302 (short time) | Cast, thermoset, grind, polish <br> Abrasion and impact resistant, high shrinkage spectacle lenses |
| 3 | 1.491 | 57.2 | Methyl methacrylate Acrylic Lucite Plexiglas | 92 | 200 | Workhorse, moldable, can be polished, good stability, shrinkage 48 hr @70°C = 1% 110°C = 6% <br> 0.2 to 0.6 % after mold cools |
| 1 | 1.590 | 30.9 | Polystyrene Lustrex Dylene Styron | 88 | 180 | In comparison to acrylic: more moldable, higher index, less stable, absorbs more water, unstable in near UV |
| 2 | 1.571 | 35.3 | Styrene acrylonitrile Lustran Tyril SAN | | | Like styrene, yellowish, low expansion coefficient Stable but tends to yellow |

Table 6.1 (Continued)

| Relative cost | n | $\nu_d$ | Common name/trade name | %Ta | °Fb | Remarks |
|---|---|---|---|---|---|---|
| 2 | 1.562 | 35 | Copolymer styrene/ acrylic NAS (30% styrene 70% acrylic) | 90 | 190 | Similar to SAN, clear |
| 4 | 1.586 | 34.7 | Polycarbonate Lexan Merlon | 89 | 255 | Excellent high temperature stability; stronger, but easily scratched; more difficult to mold and machine; use confined to injection molding; less stable; when heated to 104°C, permanent expansion = 0.7% |
|  | 1.633 |  | Polysulfone |  |  | Difficult to mold, high shrinkage, not in general use |

| 1.466 | 56.4 | Methylpentene polymer TPX | | Same as above; close to acrylic optical properties; better temperature resistance; very tough; chemically resistant; shrinkage: 0.15 to 0.30 cm/cm during molding |
| 3 | | Terpolymers of acrylonitrile, butadiene, and styrene ABS | 80 (6.35 mm) | Tough |

[a]Luminous transmittance, 0.125-in.-thick sample.
[b]Service temperature.

*Cost of tooling.* Plastic elements are cost effective only when mass produced, because of the very high cost of tooling. Small production runs are expensive, and prototype cost is generally prohibitive.

*Thermal stability.* Plastics soften at relatively low temperature, and the temperature dependence of the index of refraction is far greater than in glass.

*Scratch resistance.* Plastics are much more easily scratched than glass.

*Coating.* Antireflection and other coatings are much more difficult to apply because of thermal limits and chemical reactions.

*Material selection.* There are few materials from which to choose, and

*Optical data.* Since the total tonnage of optical plastics is small, the material suppliers do not supply much optical data on their optical grade plastics.

### 6.3  QUALITY

Plastic elements can be figured to the same tolerances as the equivalent glass. Since the molded part is the finished part, attention to mold design, finish, maintenance, and close process controls is essential. In addition, the shrinkage of the plastic (on the order of 0.002—0.006 cm/cm) must carefully be accounted for in the mold design. Scratch-dig of 60—40 or better and surface irregularities to 1/8th wave on small flats and some spherical surfaces can be maintained. Scratch-dig refers to scratches or defects in the optic element. The numerical designations are determined by visual comparison to standardized defects. These ratings are quantified in Sec. 7.3.

In multicavity molds with volume production, focal length can be held to 1%—2% and in single cavity molds held as low as 0.5%—1%. In multicavity high volume production surface figures of small flats and some spherical surfaces can be held to five fringes. For aspherics, in high volume, multicavity production of lens arrays, prolate and oblate spheroids can be controlled to eight to ten fringes.

Quality large optics up to 200 mm in diameter by 50 mm thick have successfully been produced. The use of plastics combines low cost with low weight. A 15-cm-diameter three-element lens in a plastic weighs 1.8 kg (4 lb), compared to the equivalent glass lens weight of 4.4 kg (10 lb). An important application of these large aspheric lenses is in projection television systems.

Table 6.2 Physical Properties of Optical Plastics

| Property | Acrylic (PMMA) | Polystyrene | Polycarbonate | NAS | BK7 |
|---|---|---|---|---|---|
| | A. Source: Palmer (1981)[a] | | | | |
| Index $n_d$ | 1.490 | 1.590 | 1.586 | 1.563 | 1.51680 |
| $\nu d$ | 57.2 | 30.9 | 30.3 | 34.7 | 64.17 |
| Density (g/cm$^3$) | 1.190 | 1.055 | 1.200 | 1.090 | 2.51 |
| Maximum service temp (°C) | 83 | 75 | 121 | 87 | >500 |
| Coefficient of linear thermal expansion ($10^{-5}$ cm/cm°C) | 6.20 | 5.00 | 6.75 | 5.6 | 0.71 |
| Young's modulus ($10^4$ kg/cm$^2$) | 3.02 | 3.16 | 2.43 | 3.30 | 83.1 |
| Impact strength (1 = lowest) (5 = highest) | 2 | 4 | 5 | 3 | 1 |
| Abrasion resistance (1 = lowest) (5 = highest) | 4 | 2 | 1 | 3 | 5 |
| Cost/lb (1 = lowest) (5 = highest) | 3 | 1 | 4 | 2 | 5 |

Table 6.2  (Continued)

B:  Source:  Hinchman (1980)[b]

| | Methyl methacrylate (acrylic) | Polystyrene (styrene) | Polycarbonate | Methyl methacrylate styrene copolymer |
|---|---|---|---|---|
| Specific gravity (density) | 1.19 | 1.06 | 1.20 | 1.14 |
| Refractive index (n) | 1.491 | 1.590 | 1.586 | 1.563 |
| Abbe value ($\nu$d) | 57.2 | 30.9 | 30.3 | 34.7 |
| Luminous transmittance (0.125 in. thickness) | 92% | 88% | 89% | 90% |
| Haze % | <3 | <4 | <4 | 4 |
| Critical angle | 42.2° | 39.0° | 39.1° | 39.8° |
| Service temperature limit | 200°F | 180°F | 255°F | 190°F |
| Trade names | Lucite Plexiglas | Dylene Styron | Lexan Merlon | NAS |

[a]A. L. Palmer, Practical design considerations for polymer optical systems, Contemporary Methods of Optical Fabrication, Proc. SPIE 306, 18–23, 1981.
[b]Reprinted from The Optical Industry & Systems Purchasing Directory, 1980.

## 6.4  PROPERTIES

Some physical properties of plastics used for optical components
are shown in Table 6.2.

An excellent achromat corrected for the F and C lines can be
made by combining an acrylic element with a styrene element.  In
this combination the secondary spectrum (the difference between
the F and D lines) can generally be more completely corrected than
with a glass achromat ($\lambda_D$ = 589.3 nm, $\lambda_F$ = 486.1 nm, and $\lambda_C$ =
656.3 nm).

Stresses occurring during processing induce birefringence in
plastics which exist even after careful annealing.  The index
variation is in the region of $0.06 \times 10^{-3}$ for acrylic and up to
$8 \times 10^{-3}$ for polystyrene.

Variation of index with temperature is much greater in plastics than in glass.  The dn/dT for acrylic is $8.5 \times 10^{-5}/°C$ (20
to 40°C), while the equivalent for BK7 glass is on the order of
$0.3 \times 10^{-5}/°C$.

Haze is the term applied to light scattering due to surface or
internal imperfections or inclusions.  Specifications for optical
plastic allow no more than one surface imperfection between 0.004
and 0.010 in./ft$^2$ and allow no internal bubbles or foreign inclusions
greater than 0.002 in.

## 6.5  FABRICATION OF PLASTIC OPTICS

Casting, compression or injection molding, and machining of blanks
can be used to fabricate optical plastic elements.  However, molding
is the preferred high production technique.

Casting is generally restricted to prototype, or unusually large
pieces.  However, for ophthalmic applications allyl diglycol carbonate cast blanks are used for spectacle lenses which are cast to final
figure or cast, ground, and polished.

Most plastics can be effectively machined with diamond tools.
Single point machining of optical surfaces is feasible with numeri-
cally controlled contour millers.  However, polishing is difficult,
plastics being far less amenable to abrasive polishing processes
than the glasses.

Compression molding is routinely used for Fresnel lenses and
lenticular arrays, where the molding plastic in either powder or
sheet form is pressed between heated dies.  The cycle time is long,
5–20 min, since the dies have to be heated and cooled during each
cycle.  The dies are generally made by nickel plate electroforming.

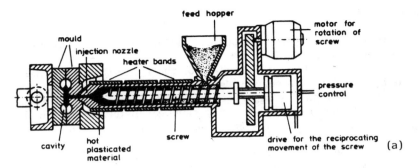

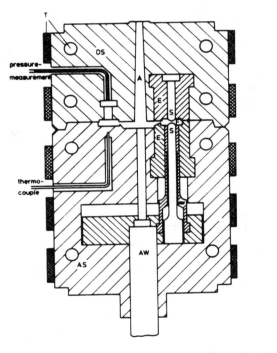

Figure 6.1   Injection molding machine—schematic (from Greis and Kirchhof, 1983).  (a) Injection machine.  (b) Mold, diagrammatic section:  DS, nozzle side; AS, ejection side; A, sprue; F, lens cavity; E, receiver, S, insert; AW, ejector; and T, temperature control.

In injection molding the molding powder is liquefied by applied heat plus mechanical work on the powder by means of a screw in a cylinder and then injected at pressures on the order of 10,000 psi into a thermally controlled mold. The plastic solidifies in the mold and is then ejected. This molding process is short, on the order of 30 sec. Often the molds are multicavity.

Injection molding is the most cost effective approach to the fabrication of plastic optical elements in large numbers. Figure 6.1 illustrates the most important elements of a molding machine. The mold consists of a base which contains the cooling passages, the sprue bushing to mate with the injection nozzle, and various movable plates to actuate the ejection pins.

The cavity within which the optical element is molded is composed of receivers containing nonoptical details of the part and polished inserts defining the optical surfaces. The receivers and inserts are usually made from hardened chromium stainless steels. These units are ground and polished by conventional glass-working techniques.

## 6.6 COATING

Antireflection coatings are routinely applied to acrylic optical elements by vacuum deposition, thus reducing surface reflection from 3.9% to <2%. Tailored multilayered coatings can reduce surface reflection to <0.5% per surface. Vacuum deposited aluminum, gold, or silver can be applied as well.

## 6.7 ADHESIVES

Epoxies and methacrylates are widely used for adhesives with index refraction varying from 1.47 to 1.61. Initially, Canada balsam ($n = 1.52$, $\nu_d = 42$) was used as a cement but was supplanted due to its temperature and humidity sensitivity. Canada balsam is made from the sap of the North American balsam fir. It is a turpentine consisting of essential oils and resins and is soluble in xylol. Polyester styrene was introduced about 1950, and today a variety of adhesives are available which are servicable between −85 and +180°F without failure.

The critical characteristics for the selection and development of an adhesive are

1.  Optical properties ($n$, $\nu_d$, absorption coefficient);
2.  Temperature stability;

Table 6.3  Optical Adhesives

|  |  | Applications | |
| --- | --- | --- | --- |
| Company | Product | Glass to glass | Glass to metal |
| Ciba-Giegy Corp.<br>Saw Mill River Rd.<br>Ardsley, N.Y. 10502 | Araldite 502 | X | |
| Dow Corning Corp.<br>Midland, Mich. 48640 | Silicone Rubber Sealer | | X |
| Eastman Kodak Co.<br>Special Product Sales<br>Rochester, N.Y. 14650 | HE-2<br>HE-63<br>HE-79<br>HE-S-1<br>HE-F-4<br>HE-10 | X<br>X<br>X<br>X<br>X | <br><br><br><br><br>X |
| Emerson & Cumming<br>Northbrook, Ill. 60062 | Stycast 1269<br>Eccoband 24<br>Ecconbond 45 | X<br>X | <br><br>X |
| Epoxy Technology<br>65 Grove St.<br>Watertown, Mass. 02172 | Epo Tek 301<br>Epo Tek 305<br>Epo Tek 310 | X<br>X<br>X | <br><br>X |
| Furane Plastics<br>P.O. Box 791<br>Rahway, N.J. 07065 | Uralane X-87718 | X | |
| General Electric Co.<br>Silicone Products Dpt.<br>Waterford, N.Y. 12188 | RTV-108 Silicone rubber<br>RTV-118 Silicone rubber<br>RTV-602<br>RTV-615<br>RTV-655 | | X<br>X<br>X<br>X<br>X |
| Narmco Materials, Inc.<br>600 Victoria St.<br>Cost Mesa, Ca. 92627 | Metlbond 2004/2102 | | X |
| Norland Products, Inc.<br>P.O. Box 145<br>North Brunswick, N.J.<br>08902 | Norland Optical Adhesive 60<br>Norland Optical Adhesive 61<br>Norland Optical Adhesive 65 | X<br>X | <br><br>X |

Table 6.3   (Continued)

|  |  | Applications | |
|  |  | Glass to glass | Glass to metal |
| Company | Product |  |  |
| Opticon Chemical | Opticon UV-57 | X | |
| P.O. Box 2445 | Opticon SFA-23 | X | |
| Palos Verdes | Opticon UVF-171 | X | |
| Peninsula, Ca. 90274 | Opticon RT-36 | | X |
|  | Opticon RT-42 | | X |
| Summers Laboratories | Lens Bond C-59 | X | |
| Morris Rd. | Lens Bond M-62 | X | |
| Fort Washington, Pa. | Lens Bond UV-69 | X | |
| 19034 | Lens Bond UV-71 | X | |
|  | Lens Bond UV-74 | X | |
| Techkits, Inc. | Epoxy Adhesive E-7 | X | X |
| P.O. Box 105 | | | |
| Demarest, N.J. 07627 | | | |
| Tra-Con, Inc. | Tra-Bond 2114 | X | X |
| 55 North St. | Tra-Bond 2113 | X | |
| Medford, Mass. 02155 | Tra-Bond 2147 | | X |
| Transene Co. | Poly-Clear | X | |
| Rte. 1 | Epoxy 20, 30 | | X |
| Rowley, Mass. 01969 | Epoxy 50 | | X |
|  | LED Clear | | X |

[a]Source:   Norland (1980).   Reprinted from *The Optical Industry and Systems Purchasing Directory*, 1980.

3.   Temperature sensitivity of optical properties;
4.   Viscosity, shrinkage, and volatiles;
5.   Separability (for rework and repair);
6.   Thermoset versus thermoplastic; and
7.   Toxicity.

A comprehensive list of commercial adhesives used for optical assemblies is provided in Table 6.3.

## REFERENCES

U. Greis and G. Kirchhof, Injection molding of plastic optics,
Proc. SPIE *381*, Bellingham, WA, 1983.

D. F. Hinchman, Developments in plastic optics, *The Optical Industry and Systems Purchasing Directory*, Optical Publishing
Co., Pittsfield, MA, 1980.

E. A. Norland, Optical cements and adhesives, *The Optical Industry and Systems Purchasing Directory*, Optical Publishing
Co., Pittsfield, MA, 1980.

A. L. Palmer, Practical design considerations for polymer optical
systems, Proc. SPIE *306*, Bellingham, WA, 1981.

Rohm and Hass Company, *Plexiglas DR-Acrylic Molding Pellets*,
Philadelphia, PA, 1980.

Rohm and Haas Company, *Technical Data PI 165g Plexiglass Acrylic
Plastic Molding Pellets*, Philadelphia, PA, 1981.

U.S. Precision Lens, Inc., *The Handbook of Plastic Optics*, Cincinnati, OH, 1983.

R. F. Weeks, Plastic optics, *Opt. News* September 1975.

H. D. Wolpert, A close look at optical plastics, *Photonics Spectra*
*17(2)*, February 1983; and *17(3)*, March 1983, Pittsfield, MA.

# 7
# OPTICAL COATINGS

## 7.1 TYPES OF COATINGS

A wide variety of optical coatings (thin films) are applied to optical elements to modify the physical, electrical, and/or optical behavior. Among the functions performed by such coatings are antireflecting (AR), beam splitting, enhanced reflection, transparent-electrically conductive, neutral density attenuation, and dichroic separation. These are listed in Table 7.1.

The ability of a coating to perform these various functions is dependent on the coating thickness, its optical characteristics, and the wavelength or bandpass of interest. A single layer antireflection coating reduces surface reflection by means of interference effects which occur between the reflected waves from the first and second surface of the film.

It is well known that the air—glass substrate interface exhibits a reflection loss for normal incidence which is given by

$$r = \frac{(n_{glass} - n_{air})^2}{(n_{glass} + n_{air})^2}$$

or approximately,

$$r = \frac{(1.5 - 1)^2}{(1.5 + 1)^2} = 0.04 \text{ or } 4\%$$

By placing a coating of thickness $\lambda/4$ with n intermediate between that of air and glass, the reflection is reduced to a value given by

$$r = \frac{(n_{film})^2 - (n_{glass})(n_{air})}{(n_{film})^2 + (n_{glass})(n_{air})}$$

For $r = 0$,

$$(n_{film})^2 = (n_{air})(n_{glass}) \quad \text{or} \quad n_{film} = [(n_{air})(n_{glass})]^{\frac{1}{2}}$$

Table 7.1  Types of Optical Coatings

| Type | Function |
|------|----------|
| Antireflection (AR) | |
|   V-Coats | Selective wavelength antireflection |
|   Broad band | Broad band antireflection |
| Beam splitters | Partial mirrors |
| Mirrors | First surface reflector<br>Second surface reflector<br>  (dielectrics or metal films) |
| Conductive—transparent | Radio frequency interference<br>Antistatic |
| Filter | Narrow band<br>Broad band<br>Neutral density |
| Dichroics | Heat/light separation<br>Color separation<br>Bandpass |

Usually, for an optical element in the visible bandpass, a yellow green is chosen leaving some residual reflection in the violet and red which make such coatings look purple.

Considerable improvement over the typical one-layer coating is possible by the use of multilayer coatings. A similar concept is employed for making dichroic mirrors. A typical single layer AR coating will be effective over a spectral range of about 2, i.e. from 2/3 to 3/2 of the principal wavelength. A single layer AR coating on crown glass will be effective from about 400 to beyond 750 nm.

Since the optical path length in the film varies with the angle of incidence, to obtain high performance over a wide range of angular incidence requires multilayer coatings.

Single layer AR coatings are used when there are not many optical elements in the system and the preservation of the initial light level is not critical. These are relatively broad band, relatively insensitive to angle of incidence and comparatively low cost. A typical single layer AR coating spectral curve is shown in Fig. 7.1.

Broad band AR coatings require multiple layers of dielectric materials and are used where efficiency of light transmission over a broad spectrum is needed. These coatings are more sensitive to

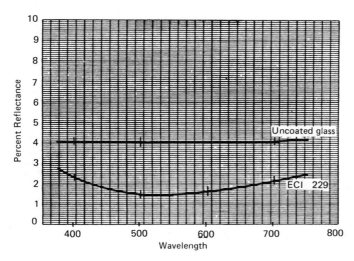

**Figure 7.1**  Single layer AR coating—spectral reflectance (from Evaporated Coatings, Inc., 1983).  ECI No. 229:  Single layer coating of magnesium fluoride—this coating is in conformance with MIL-C-675A and can be shifted to provide a minimum reflectance of under 1.5% at any specified point in the visible or near UV or IR regions.

angle of incidence.  A typical transmission spectrum of such a coating configuration is indicated in Fig. 7.2.
"V-coats" or narrow band multilayer coatings are optimized for a narrow frequency range.  These are also sensitive to angle of incidence (Fig. 7.3).

Second surface reflections are suppressed by coating the optical element on both sides.  Additional reflection and transmission curves for a variety of generic coating types are shown in Figs. 7.4 to 7.7.

## 7.2  COATING PROCESSES

Conventional coating processes depend on evaporative supply of the coating material from a source and subsequent condensation of the evaporant on the cool substrate.  Evaporation can be accomplished by simple electrical heating or for refractory (high melting temperature) materials such as $Al_2O_3$ ($T_{mp}$ = 2030°C) electron beam guns are used.

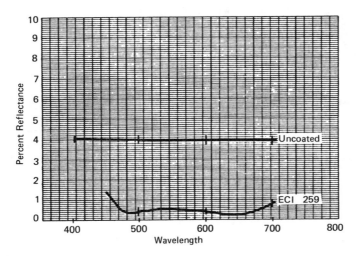

**Figure 7.2** Broad band AR coating—spectral reflectance (from Evaporated Coatings, Inc., 1983). ECI No. 259: Multilayer low reflection coating provides 1% average reflection across the visible spectrum.

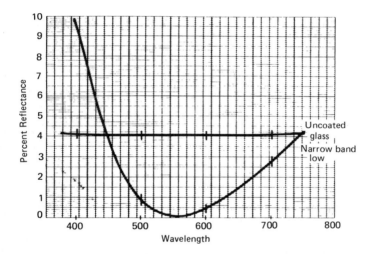

**Figure 7.3**   V-coat narrow band—spectral reflectance (from Evaporated Coatings, Inc., 1983).   ECI No. 269:   V-coat R $\leqslant$ 0.2% at $\lambda_0$.

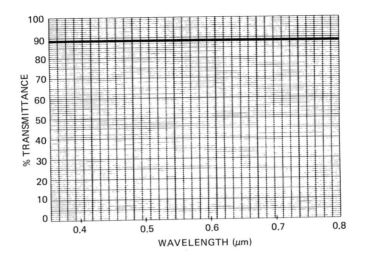

*Description*  The ECI 1258 permanent high transparency anti-static coating combines permanent static shielding with high visible light transmission.  Applied by high vacuum evaporation, permanency remains unaffected by changes in temperature, humidity, or other conditions.  The coating is available in standard ranges from 0.1 megohms per square surface resistivity with 82% visible light transmission to 10 megohms per square with 88% light transmission. Neutral in color, the ECI 1258 permanent antistatic coating is virtually undetectable.  If desired, it may be combined with antireflection or other coatings to provide even greater light transmission or other desirable control functions.

*Applications*  ECI Antistatic Coating 1258 is espectially suitable for static shielding of meter and instrument face plates without impairment visibility.  The coating prevents false or erratic readings caused by static buildup.  It can be applied to glass, ceramic, and most plastics and is suitable wherever permanent antistatic shielding is required.

Figure 7.4    Antistatic coating for glass and plastic—spectral transmittance (from Evaporated Coatings, Inc., 1983).

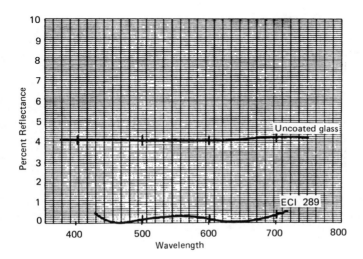

**Figure 7.5**   Broad band AR—spectral reflectance (from Evaporated Coatings, Inc., 1983).   ECI No. 289:   High efficiency broad-band low reflection coating—this coating is in conformance with MIL-C-14806A.   R ⩽ 0.5% from 440—630 nm.

*Typical characteristics*

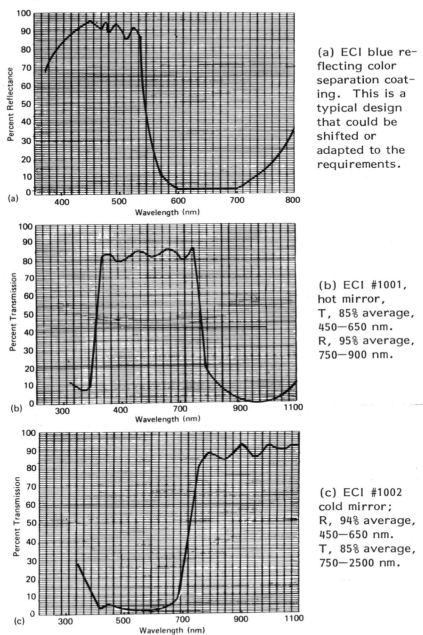

(a) ECI blue reflecting color separation coating. This is a typical design that could be shifted or adapted to the requirements.

(b) ECI #1001, hot mirror, T, 85% average, 450—650 nm. R, 95% average, 750—900 nm.

(c) ECI #1002 cold mirror; R, 94% average, 450—650 nm. T, 85% average, 750—2500 nm.

**Figure 7.6** Color separation and heat light separation coatings—spectral transmittance (from Evaporated Coatings, Inc., 1983).

The coating system includes the evaporator means, vacuum and gas systems for maintaining the appropriate chamber environment, glow discharge system for substrate cleaning, substrate heater, rotational work holders, and monitors for determining the thickness of the film. A typical simple vacuum system is illustrated in Fig. 7.8.

Many coating materials can be used depending on the desired effects including Cr, Au, Al, Ag, Inconel, Rh, Ti, and Ge, among the metals; and $CeO_2$, $MgF_2$, $ThF_4$, quartz ($SiO_2$), $Al_2O_3$, $ZrO_2$, ZnS, and $CeF_3$, among the nonmetals. Dielectric stacks (multiple layers of different dielectric materials) are frequently used.

The coating material chosen must be appropriate for the desired wavelength region, compatible with other adjacent coating materials and substrate, capable of meeting environmental exposures and stresses, and be relatively easy to deposit. In addition, it is advantageous if the coating is removable for rework, when necessary.

The surface must be "clean" prior to starting deposition. A typical cleaning procedure for a glass consists of the following steps: clean with precipitated chalk and water, rinse, solvent wipe, and vapor degrease. The specific cleaning procedure must be selected in consideration of the surface hardness, chemical solubilities, and staining characteristics of the substrate.

---

**Figure 7.6**   (Continued)

*Description*  Color separating or dichroic filter coatings are used to split an incident beam into two beams of different spectral color. ECI offers a tremendous variety of coating designs or a design can be generated for your particular application. A particular coating design could be applied to a special absorbing substrate material to meet your exact spectral requirements. ECI #1001 heat reflecting and visible light transmitting (hot mirror) coating design is used to separate unwanted heat from an optical system. ECI #1002 visible light reflecting and heat transmitting (cold mirror) coating design is used as a more efficient means of removing heat because a wider range of infrared wavelengths is affected. All of these designs are multilayer, dielectric designs that are extremely durable and temperature resistant.

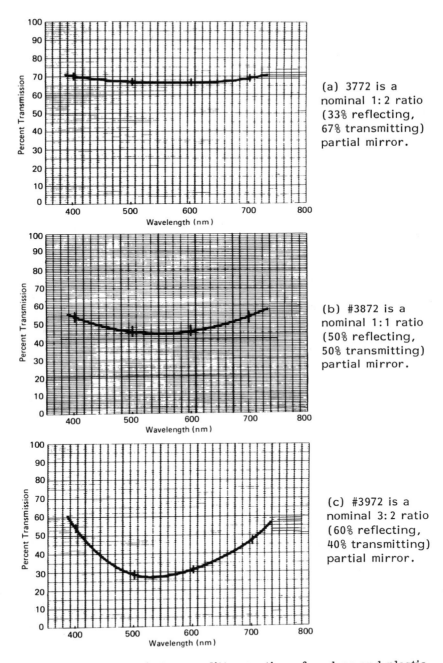

(a) 3772 is a nominal 1:2 ratio (33% reflecting, 67% transmitting) partial mirror.

(b) #3872 is a nominal 1:1 ratio (50% reflecting, 50% transmitting) partial mirror.

(c) #3972 is a nominal 3:2 ratio (60% reflecting, 40% transmitting) partial mirror.

**Figure 7.7** Dielectric beam splitter coatings for glass and plastic optical elements—spectral transmittance (from Evaporated Coatings, Inc., 1983).

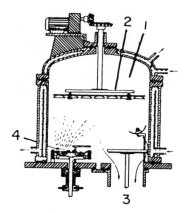

**Figure 7.8** Simple vacuum coating system (from Feldman, 1980). Reprinted from *The Optical Industry and Systems Purchasing Directory*, 1980. A schematic of a typical vacuum deposition system: (1) Metal or glass vacuum chamber (bell jar); (2) the support for the objects being coated; (3) system for production and maintenance of the vacuum (not shown here); and (4) illustrates an evaporation device.

**Figure 7.7** (Continued)

*General description* A variety of neutral, highly efficient, nonabsorbing partial transmitting mirrors are available. The curves shown are for typical R:T ratios; however, almost any ratio is available. These coatings may be used as first surface beam-splitters or cemented in cubes. The coating can be designed for polarized or unpolarized light and for various angles of incidence.

Evaporation sources include:

Resistance heating up to 1800°C, e.g., $MgF_2$, $SiO_2$, $ZnSO_4$.
Sources may be helical coils which evaporate or material may be
held in heated boats.
Electron-beam from 1800–3500°C, e.g., $CeO_2$, $TiO_2$, $ZrO_2$, $Ta_2O_5$.
In this process the container is water cooled to prevent thermal
reactions and meltdown of the container. The E-beam is focused
to produce heat concentration at the focal point.

Cathode sputtering, another made for activating the source,
uses electrostatic acceleration of Ar ions to bombard the source.
This technique is capable of yielding very high purity and very
adherent coatings, and can deposit very refractory materials.

Thickness monitoring can be accomplished optically by the use
of a calibrated witness plate as indicated in Fig. 7.9 by monitoring
either or both reflection and transmission through the plate. The
optical monitor operates by indicating maxima and minima of reflectance and transmittance. Each extremum represents one quarter
wave, $\lambda/4$, of optical thickness.

In the glow discharge surface cleaning process, electric potential is designed to deflect electrons and allow positive ions and neutral molecules to strike the work which clean the surface. The high
energy electrons must be deflected, since they induce contamination on the surface.

In evacuating the apparatus before deposition, glow discharge
cleaning takes place at about $10^{-4}$ to $10^{-5}$ Torr. The source is
then heated to the evaporation temperature to initiate coating of
the substrate.

More elaborate coating systems are indicated in Fig. 7.10 to
7.12. The planetary substrate fixturing indicated in Fig. 7.10 was
devised to enhance uniformity of deposition on large pieces or on
many small pieces being coated simultaneously.

A drum arrangement for very large production runs is indicated
in Fig. 7.11; a continuous coater designed to lay down multilayer
coatings is shown in Fig. 7.12. As the machine size increases, the
attainable precision of deposition suffers to some extent. Table 7.2
shows the relationship where "Machine size" refers to the inside
horizontal dimensions of a square or round chamber, and "Thickness
control" is the total variation across the part, part-to-part, and
between different coating runs.

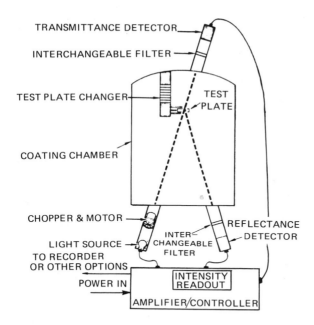

**Figure 7.9** Thickness monitoring setup (from Feldman, 1980). Reprinted from *The Optical Industry and Systems Purchasing Directory*, 1980. Basic layout of an optical thin film thickness monitor; light from the source is reflected or transmitted at different intensities depending upon the optical thickness of the coating being deposited on the test plate. This transmittance or reflectance is detected at the other end of the optical system and displayed by the control unit.

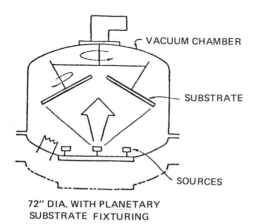

**Figure 7.10** Planetary coater (from Seddon, 1978).

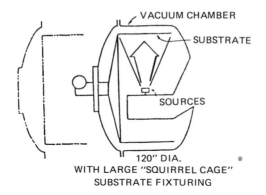

VACUUM CHAMBER

SUBSTRATE

SOURCES

120" DIA.
WITH LARGE "SQUIRREL CAGE"
SUBSTRATE FIXTURING

Figure 7.11   Squirrel cage coater (from Seddon, 1978).

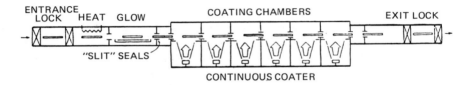

ENTRANCE
LOCK   HEAT   GLOW              COATING CHAMBERS                    EXIT LOCK

"SLIT" SEALS

CONTINUOUS COATER

Figure 7.12   Continuous coater (from Seddon, 1978).

Table 7.2   Attainable Precision versus Machine Size

| Machine size (in.)[a] | Substrate size (max.) (in.) | Thickness control (%)[b] |
|---|---|---|
| 30 | 2 × 3 | ±(0.1−0.2) |
| 48 | 5 × 7 | ±(0.2−0.4) |
| 84 | 11 × 17 | ±(0.5−1.0) |

[a]Machine size refers to the inside horizontal dimensions of a square or round chamber.
[b]Thickness control is total for variation across part, part-to-part, and between different coating runs.
Source:  Seddon, 1978.

## 7.3 SURFACE QUALITY

Coatings are tested for compliance with humidity resistance, coating adhesion, and scratch-dig requirements. Tests for these properties are standardized.

The scratch number is the apparent width of a hairline scratch specified in units of 0.001 mm (1 μm). Thus, a number 60 scratch is equivalent to a scratch of 60 × 0.001 or 0.06 mm width.

Dig numbers relate to point defects such as bubbles, pinholes, and inclusions. Digs are specified in units of 0.01 mm so that a number 60 dig represents a point defect of 0.6 mm diameter.

## REFERENCES

Evaporated Coatings, Inc., *Coating Data Sheets*, Huntingdon Valley, PA, 1983.

J. Feldman, Vacuum equipment for optics, *Optical Industry and Systems Purchasing Directory*, The Optical Publishing Co., Inc., Pittsfield, MA, 1980.

I. Seddon, Opportunities in Thin Films to Meet Energy Needs, *Optical Engineering 17*, No. 5, (1978).

# 8
# FILTERS

## 8.1  INTRODUCTION

An optical filter attenuates particular wavelengths of radiation while passing others with relatively little change.  The attenuation can be effected by absorption in the body of the filter (absorption filter) or interference effects in a single or multilayer coating (interference filter).  Neutral density filters attenuate light without changing the spectral quality of the light.  An excellent summary of the various types of filters which have been developed is given by Dubrowolski (1978).

## 8.2  ABSORPTION FILTERS

Absorption filters can rely on selective scattering or on bulk absorption as described by Bouguer's law

$$I = I_0 e^{-\alpha t}$$

where $\alpha$ is the absorption coefficient ($cm^{-1}$), $I_0$ is the incident beam intensity, and $I$ is the exit beam intensity.

A spectral filter is a device which selects specific wavelengths for absorption with high precision.  A major application of absorption filters is the selective absorption of certain band passes in the visible spectrum, or a color filter.

Selective scattering filters operate by inclusion of closely packed microparticles or microcrystals in a transparent host.  The materials are so chosen that the index of refraction of the particle and the host are the same at the wavelength to be transmitted.  The two materials are chosen with widely different dispersion, $dn/d\lambda$, characteristics.  As the wavelength departs from the equi-index wavelength, scattering increases in accordance with Rayleigh (particle size $\ll \lambda$) or Mie (particle size $\cong \lambda$) scattering theory.  Quartz powder in liquids or gases have been used (Christiansen filter) as indicated in Fig. 8.1. The center of the passband can be varied by changing the temperature.

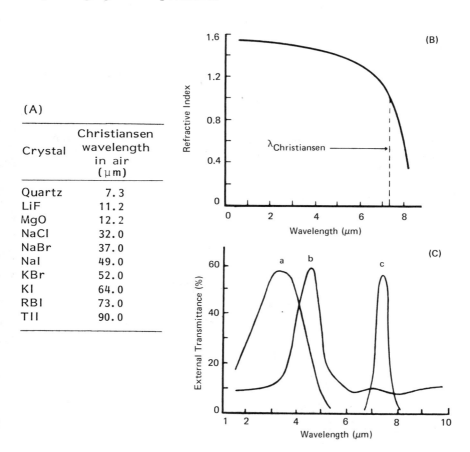

(A)

| Crystal | Christiansen wavelength in air ($\mu$m) |
|---------|-------------------------------------|
| Quartz | 7.3 |
| LiF | 11.2 |
| MgO | 12.2 |
| NaCl | 32.0 |
| NaBr | 37.0 |
| NaI | 49.0 |
| KBr | 52.0 |
| KI | 64.0 |
| RBI | 73.0 |
| TlI | 90.0 |

Figure 8.1 Christiansen filter (from Wolfe and Zissis, 1982). (A) Christiansen wavelengths of selected materials. (B) Dispersion curve of quartz, showing Christiansen wavelength in air. (C) Position of the Christiansen peak for quartz powder in liquids. (a) Quartz in a 50% by volume mixture of $CS_2$ and $CCl_4$. (b) Quartz in pure $CCl_4$. (c) Quartz in air. Reprinted with permission of W. L. Wolfe and G. J. Zissis, *The Infrared Handbook*, Environmental Research Institute of Michigan, Ann Arbor, MI, 1982.

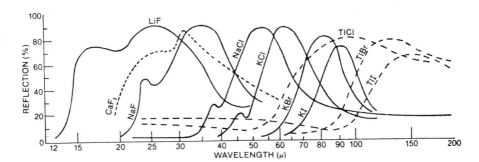

**Figure 8.2**    Reststrahlen reflection of binary halides (from Harshaw, 1982).   Reprinted by permission of Harshaw/Filtrol
Partnership.

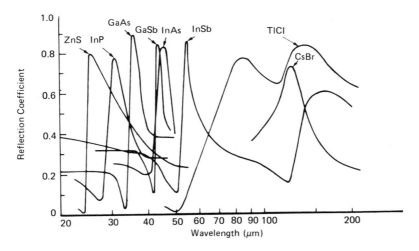

**Figure 8.3**    Reststrahlen reflection of some crystalline substances
(from Wolfe and Zissis, 1978).   Reprinted with permission of W. L.
Wolfe and G. J. Zissis, *The Infrared Handbook*, Environmental
Research Institute of Michigan, Ann Arbor, MI, 1978.

Selective reflection filters depend on the reststrahlen reflection of certain crystals. These reflections are shown in Figs. 8.2 and 8.3 for alkali halides and for some other substances. Radiation is collimated and directed at a polished surface of a crystal which has its reststrahlen reflection peak in the band pass of interest. After three or four similar reflections the reststrahlen wavelength is predominant, the other wavelengths having been attenuated by orders of magnitude.

A common scattering technique is to precipitate a second phase particle in a glass host through heat treatment (striking process). Since for particles with diameter $\ll \lambda$, scattering is proportional to $\lambda^{-4}$, the shorter wavelengths are scattered and the higher wavelengths are transmitted. Figure 8.4 illustrates the sharp cut-on glass filters for the visible region. The more narrow the particle size distribution, the sharper is the cut-on. The total number of scatterers per unit volume also affects the filtering performance. An example of this type of filter is one

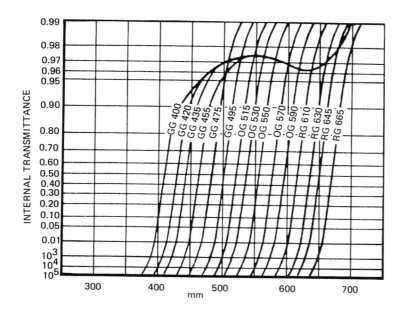

**Figure 8.4** Sharp cut-on scattering glass filters (from Schott Glass Technologies, Inc., 1982). Reprinted by permission of Schott Glass Technologies, Inc.

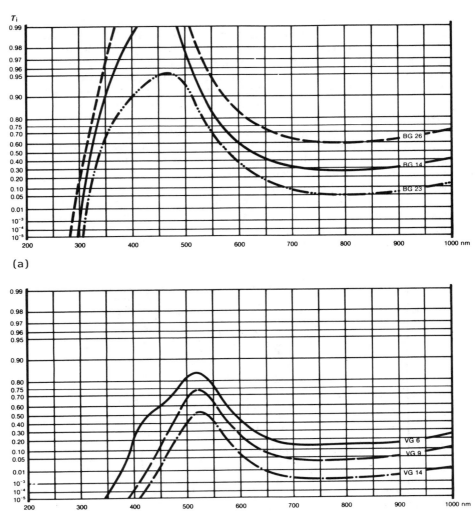

**Figure 8.5** Absorption filters (from Schott Glass Technologies, Inc., 1982). (a) Blue transmitting, red absorbing filters. (b) Green transmitting, red/blue absorbing filters. Reprinted by permission of Schott Glass Technologies, Inc.

**Table 8.1    Characteristics of Absorbing and Scattering Filters[a]**

|  | Ionic coloration types | Crystallite types |
|---|---|---|
| Production steps: | Batching, melt, forming | Batching, melt, forming with subsequent thermal treatment |
| Colorant mechanisms: | Simple absorption by ions in true solution | Absorption by microcrystals predominates |
| Typical colorants: | Nickel oxide (purple) Cobalt oxide (blue) Chromium oxide (green) | Sulfur (light yellow) Cadmium sulfide (yellow) Cadmium selenide (orange-red) Gold (rose to ruby) |
| Shape of the spectral curve depends on: | Quantity and type of colorants. Relative proportions. Oxidation state. Host glass composition. Glass thickness | Quantity and type of colorants. Relative proportions. Post production thermal treatment. Host glass composition. Glass thickness |
| Shape of the spectral curve: | Sine- or bell-curve | Steeply rising cut-on, long wave region of high transmission and short wave, strongly absorbing region |
| Spectral range of application: | Number of typical curves limited by availability of suitable absorbing ions | Cut-on points may be located at any wavelength in the region from 400 to 700 , limited number of other types available in higher and lower regions |
| Deviations can occur: | From melt to melt as a result of minute variations in properties and purity of the raw materials | Within a melt, and to a small extent within a block, because of minute variations in the effective temperature cycle |

[a]Source:  Schott Glass Technologies, Inc. (1981).  Reprinted by permission of Schott Glass Technologies, Inc.

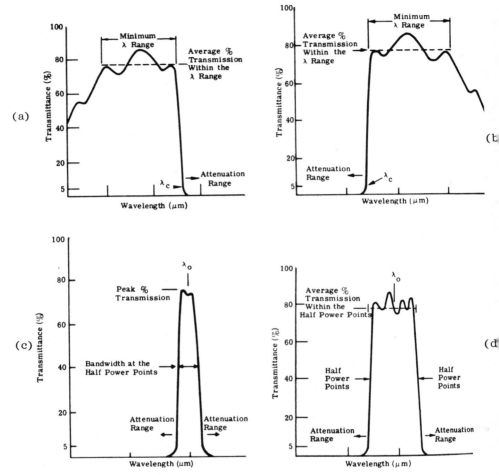

**Figure 8.6**  Spectra for filter glasses (from Wolfe and Zissis, 1978).
(a) Short wavelength pass filter.  (b) Long wavelength pass filter.
(c) Narrow bandpass filter.  (d) Wide bandpass filter.  Reprinted
with permission of W. L. Wolfe and G. J. Zissis, *The Infrared
Handbook*, Environmental Research Institute of Michigan, Ann
Arbor, MI, 1978.

in which gold is dispersed in a glass host to create a ruby col-
ored glass.

Gelatin host filters employ a wide variety of dyes to control the
band pass spectral position.

Absorption filters are produced in a variety of host materials
such as gelatin, glass, liquid, or plastic where the absorbing component actually absorbs the radiation. These types of filters have
a more gradual cut-on and cut-off than the scattering type, as
indicated in Fig. 8.5. These filters have the advantage of reject-
ing both higher and lower wavelength radiation than the band pass
of interest.

A summary of key characteristics of the absorbing (ionic colora-
tion) type and the selective scattering type filter is presented in
Table 8.1. Representative spectra for a variety of filter glasses
are shown in Fig. 8.6.

## 8.3 INTERFERENCE FILTERS

The interference filter depends on multilayer dielectric-metal stacks
to control the transmission or reflection of particular bandpasses
by means of constructive and destructive interference. The design
of such stacks is beyond the scope of this volume but is well cov-
ered by Dobrowolski (1978) and in other texts.

Common substrates and commonly used films are given in Tables
8.2 and 8.3. Custom designed filters to perform a wide variety
of optical tasks can be obtained. Typical interference filter spectra
are illustrated in Fig. 8.7.

## 8.4 NEUTRAL DENSITY FILTERS

The neutral density filter attenuates light without changing its
spectral quality. One way of specifying a neutral filter is by optical density (OD), where OD is defined by the following relation:

$$OD = \log_{10} 1/T$$

where T is the external transmittance. If OD is specified as 1.0,
then T = 10%.

One type of neutral filter is composed of an evaporated metal
such as Cr, Ni, or an alloy such as nichrome or monel deposited
on a substrate such as glass, quartz, or other optical transmitting
material. Such filters have little scattering and are neutral from

Table 8.2   Common Substrates[a]

| Material | Refractive index | Transmission range | Material | Refractive index | Transmission range |
|---|---|---|---|---|---|
| Irtran[b] 1 | 1.38–1.23 | 1.00–9.00 | Magnesium oxide | 1.77–1.62 | 0.36–5.35 |
| Lithium fluoride | 1.45–1.11 | 0.20–9.80 | Sapphire | 1.83–1.59 | 0.27–5.60 |
| Calcium fluoride (also as Irtran 3) | 1.44–1.32 | 0.20–12.00 | Irtran 2 | 2.29–2.15 | 1.00–13.00 |
| | | | Irtran 4 | 2.50–2.30 | 1.00–20.00 |
| Vycor[c] | 1.46 | 0.25–3.50 | Arsenic trisulfide | | |
| Fused quartz | 1.48–1.41 | 0.20–4.50 | glass | 2.69–2.36 | 0.56–12.00 |
| Barium fluoride | 1.51–1.40 | 0.26–10.35 | Silicon | 3.50–3.42 | 1.36–7.00 |
| Glass | 1.70–1.51 | 0.32–2.50 | Germanium | 4.10–4.00 | 1.80–23.00 |

[a]Source: Wolfe and Zissis (1978).   Reprinted with permission of W. L. Wolfe and G. J. Zissis, *The Infrared Handbook*, Environmental Research Institute of Michigan, Ann Arbor, MI, 1978.
[b]Irtran is a registered trademark of Eastman Kodak Co.
[c]Vycor is a registered trademark of Corning Glass Works.

Table 8.3    Commonly Used Films[a]

| Material | Refractive index | Range of transparency[b] from (nm) | to (μm) | Comments |
|---|---|---|---|---|
| Cryolite | 1.35 | < 200 | 10 | c |
| Chiolite | 1.35 | < 200 | 10 | c |
| Magnesium fluoride | 1.38 | 230 | 5 | d,e |
| Thorium fluoride | 1.45 | < 200 | 10 | − |
| Cerium fluoride | 1.62 | 300 | > 5 | f |
| Silicon monoxide | 1.45−1.90 | 350 | 8 | g |
| Sodium chloride | 1.54 | 180 | > 15 | h |
| Zirconium dioxide | 2.10 | 300 | > 7 | d |
| Zinc sulfide | 2.30 | 400 | 14 | i |
| Titanium dioxide | 2.40−2.90 | 400 | > 7 | j |
| Cerium dioxide | 2.30 | 400 | 5 | d,e |
| Silicon | 3.50 | 900 | 8 | − |
| Germanium | 3.80−4.20 | 1400 | > 20 | − |
| Lead telluride | 5.10 | 3900 | > 20 | − |

[a]Source: Wolfe and Zissis (1978). Reprinted with permission of W. L. Wolfe and G. J. Zissis, *The Infrared Handbook*, Environmental Research Institute of Michigan, Ann Arbor, MI, 1978.

[b]The range of transparency is for a film of quarter-wave optical thickness at this wavelength. These values are approximate and also depend quite markedly on the conditions in the vacuum during the evaporation of the film.

[c]Both materials are sodium−aluminum fluoride compounds, but differ in the ratio of Na to Al and have different crystal structure. Chiolite is preferable in the infrared, beacuse it has less stress than cryolite.

[d]These materials are hard and durable, especially when evaporated onto a hot substrate.

[e]The long wavelength is limited by the fact that, when the optical thickness of the film is a quarter-wave at 5 μm, the film cracks because of the mechanical stress.

[f]Other fluorides and oxides of rare earths have refractive indices in this range, from 1.60 to 2.0.

[g]The refractive index of $SiO_x$ (called silicon monoxide) can vary from 1.45 to 1.90 depending on the partial pressure of oxygen during the evaporation. Films with a refractive index of 1.75 and higher absorb at wavelengths below 500 nm.

[h]Sodium chloride is used in interference filters out to a wavelength of 20 μm. It has very little stress.

[i]The refractive index of zinc sulfide is dispersive.

[j]The refractive index of $TiO_2$ rises sharply in the blue spectral region.

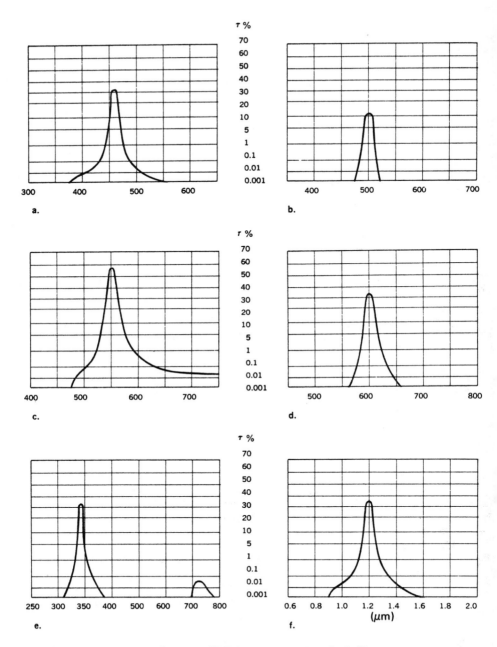

**Figure 8.7** Interference filter spectra: Typical filter curves (from Schott Glass Technologies, Inc., 1972). Reprinted by permission of Schott Glass Technologies, Inc.

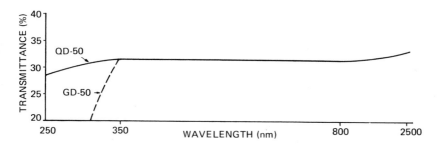

Figure 8.8   Transmittance spectra for inconel coated crown glass (GD-50) and fused silica (QD-50) (from Corion, 1983).

the UV to the IR, generally limited by the characteristics of the substrate. An example of transmittance spectra for inconel coated crown glass and fused silica are shown in Fig. 8.8.

A second type of neutral density filter consists of carbon particles suspended in gelatin. Scattering is quite high but optical density can be quite closely controlled in the visible region.

## REFERENCES

Corion Corp., *Optical Filters and Coatings*, Holliston, MA, 1983.

J. A. Dobrowolski, Coatings and filters, *Handbook of Optics*, McGraw-Hill, Inc., New York, 1978.

The Harshaw Chemical Co., *Crystal Optics*, Catalog OP 101, Solon, OH, 1982.

W. D. Kingery, H. K. Bowen, and D. R. Uhlmann, *Introduction to Ceramics*, John Wiley and Sons, Inc., New York, 1976.

Schott Glass Technologies, Inc., *Interference Filters*, Duryea, PA, 1972.

Schott Glass Technologies, Inc., *Color Filter Glass*, Duryea, PA, 1982.

W. L. Wolfe and G. F. Zissis, *The Infrared Handbook*, Environmental Research Institute of Michigan, Ann Arbor, MI, 1978.

# 9
# MIRRORS

## 9.1 SUBSTRATES

Substrates for precision mirrors must be dimensionally stable over time, have low or near zero thermal expansion over the service temperature range, be structurally stiff with low mass, have high thermal diffusivity to minimize thermal gradients and associated distortion, be capable of achieving a very smooth surface finish and of accepting a reflective coating which meets the optical requirements.

Any specific application will involve a trade-off analysis among the various substrate material candidates. Common mirror substrates are listed in Table 9.1 along with thermal, mechanical, and physical properties at 300 K. The desired properties include high stiffness to mass ratio to minimize deformation due to positioning, a high thermal diffusivity, which implies a high thermal conductivity, a low mass density, and a low specific heat. A high thermal diffusivity reduces transient temperature gradients in the mirror substrate and the consequent distortions due to nonuniform thermal expansion. Thermal conductivities of metals tend to decline and those of fused silica and silica-based glasses to rise as the temperature goes up. Thermal conductivities of some metals and nonmetals are given in Fig. 9.1a,b.

As already mentioned, a low coefficient of thermal expansion at the temperature range of operation to minimize dimensional change due to thermal growth is essential. In addition, a high microyield strength is needed so that all strains in service are totally elastic. Temporal dimensional stability is related to the processing history of the material since the materials "remember" their past and will try to relieve any residual mechanical, electrical, or chemical stresses or strains over time.

Table 9.1  Properties of Mirror Substrate Materials at 300 K[a]

| Material | ρ Density (g/cm³) | E Modulus of elasticity (10⁶N/cm²) | k Thermal conductivity (cal/cm sec °C) | c Specific heat (cal/g °C) | α Coefficient of expansion (10⁴/°C) | Specific stiffness E/ρ (10⁶ cm) | Thermal distortion index (α/D × 10⁶) | Thermal diffusivity (k/ρc) | Microyield strength (psi)[b] | Transform. temp. (°C) |
|---|---|---|---|---|---|---|---|---|---|---|
| Fused silica | 2.2 | 7.0 | 0.0033 | 0.188 | 0.55 | 3.18 | 69 | 0.008 | 1500 | 1500 |
| ULE | 2.21 | 6.74 | 0.0031 | 0.183 | 0.03 | 3.05 | 4 | 0.008 | 1500 | 1500 |
| Cer-Vit | 2.5 | 9.23 | 0.004 | 0.217 | 0.1 | 3.7 | 14 | 0.008 | 1500 | 800 |
| Aluminum | 2.70 | 6.9 | 0.53 | 0.215 | 23.9 | 2.56 | 26 | 0.92 | 2-8000 | 660 |
| Copper | 8.96 | 12.0 | 0.94 | 0.09 | 16.5 | 1.34 | 14.5 | 1.14 | 500 | 1080 |
| Beryllium | 1.83 | 28.0 | 0.38 | 0.45 | 12.4 | 15.4 | 27 | 0.46 | 2-10000 | 1250 |
| Molybdenum | 10.2 | 32.0 | 0.35 | 0.06 | 4.9 | 3.14 | 8.7 | 0.56 | 5000 | 2610 |
| Silicon | 2.33 | 13.1 | 0.31 | 0.13 | 4.2 | 5.6 | 4.0 | 1.0 | N/A | 1410 |

[a]Source: Slomba (1980). Reprinted from *Optical Industry and Systems Purchasing Directory*, 1980.
[b]Microyield strength is a permanent distortion parameter.

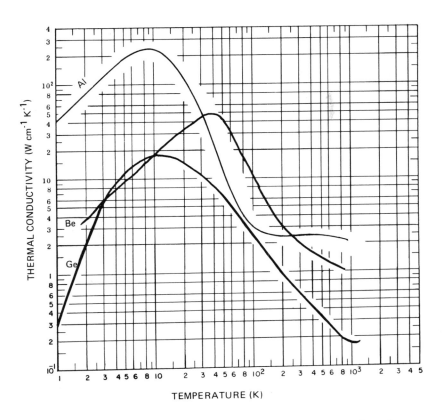

**Figure 9.1a**    Thermal conductivity of metals and a semiconductor, Al, Be, and Ge (from Touloukian et al., 1970).

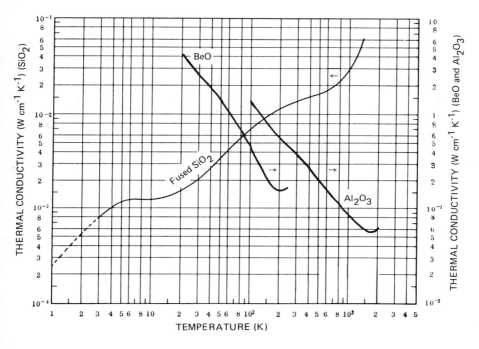

**Figure 9.1b**    Thermal conductivity of nonmetallic materials, fused silica (SiO$_2$), BeO, and Al$_2$O$_3$ (from Touloukian et al., 1970).

Figure 9.2a shows the coefficient of thermal expansion (CTE) for various substrates at 300 K. In addition, the temperature at which a zero CTE occurs for each material is listed. Whereas ultra-low-expansion ULE, Cer-Vit, and Zerodur have near zero CTE at $\sim$300 K, the absolute value of the CTE rises both below and above 300 K. Fused quartz has a zero CTE at 180 K, Be at 40 K, and Al at 15 K. Figure 9.2b compares longitudinal dimensional change of Zerodur and fused silica over the temperature range −273°C to +150°C.

| Temperature at which CTE = 0 K | Material | Coefficient of thermal expansion at 300 K $°C^{-1}$ |
|---|---|---|
| 293 | ULE | 0.0 |
|  | CERVIT | 0.1 |
| 423 | ZERODUR | −0.05 |
| 190, 273 | f $SiO_2$ (fused quartz) | 0.55 |
| 40 | Be | 12.4 |
| 15 | Al | 23.9 |

Figure 9.2a    Mirror substrates—coefficients of thermal expansion (CTE) at various temperatures.

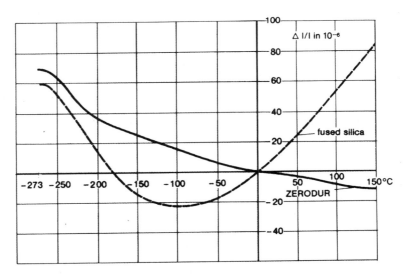

Figure 9.2b    Linear thermal expansion. Typical longitudinal change in ZERODUR relative to 0°C within the temperature range −271°C to +150°C in comparison to fused silica (from Schott Optical Technologies, Inc., 1982). Reprinted by permission of Schott Optical Technologies, Inc.

## 9.2 SURFACE FINISHING

The best possible surface finish varies with material as noted in
Table 9.2 and scattering is a direct function of surface roughness
(Fig. 9.3). Superpolishing is a new technique which is a chemo-
mechanical process capable of removing single atoms or molecules
in a controlled manner, Namba (1982). The principle of this process (called float polishing) is to remove atoms form the topmost
surface of the sample through collision with fine particles in a
polishing fluid.

Table 9.2  Typical RMS Roughness Values for Various Substrate
Materials[a]

| Material | Typical roughness (Å) | Best value observed (Å) |
|---|---|---|
| Dielectrics | | |
| Fused-quartz | 13 | 2 |
| ULE quartz | 13 | 10 |
| Dense flint | 35 | 33 |
| $CaF_2$ | 42 | 10 |
| $Al_2O_3$ | 61 | 10 |
| KCl | 110 | 29 |
| SiC | 12 | 7 |
| Metals | | |
| Al | 53 | 19 |
| Cu (sputtered) | 15 | 11 |
| (bulk) | 30 | 15 |
| BeCu | 68 | 28 |
| Mo (sputtered) | 12 | 12 |
| (bulk) | 47 | 15 |
| (TZM) | 40 | 15 |
| Invar | 47 | 28 |
| Stainless steel | 40 | 33 |
| Ti | 27 | 13 |
| Monel | 56 | 29 |
| Amorphous nickel | 18 | 11 |

[a]Source: Bennett (1978). H. Bennett, Optical properties of
optical materials, *17*(5), 1978.

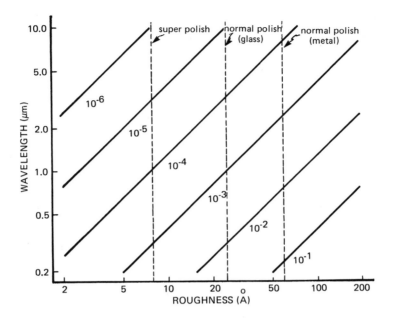

**Figure 9.3** Scattered light levels predicted theoretically (diagonal lines) for surfaces having rms roughnesses from 2 to 200 Å. The dashed lines indicate typical roughnesses of various kinds of polished surfaces. Wavelengths from the ultraviolet to the infrared are plotted logarithmically on the ordinate (from Bennett, 1978). H. Bennett, Optical properties of optical materials, Opt. Eng. *17*(5), 1978.

## 9.3  COATINGS FOR MIRRORS

Mirror substrates may be coated with metals, dielectric multi-layers, or metals with nonmetallic overcoats depending on the specific requirements.

The reflectances of conventionally used metallic coatings are shown in Fig. 9.4, and of some commercially available mirror coatings in Fig. 9.5. Reflectance data for a UV enhanced aluminum coating and for a rhodium coating are displayed in Figs. 9.6 and 9.7, respectively. Table 9.3 lists some of these coatings and their conventional uses.

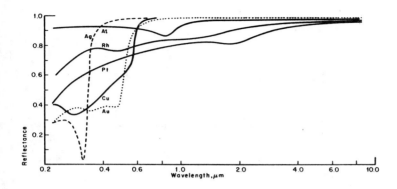

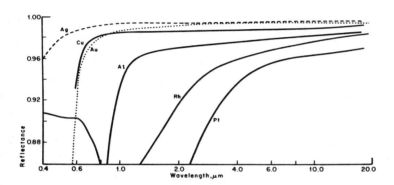

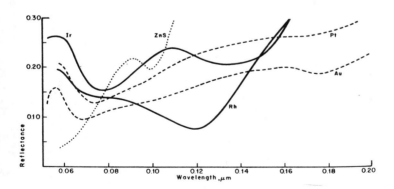

Figure 9.4 Reflectance of metallic reflectors (from Dobrowolski, 1978). Reprinted with permission of W. G. Driscoll and W. Vaushan, *Handbook of Optics*, McGraw-Hill Book Company, New York, 1978.

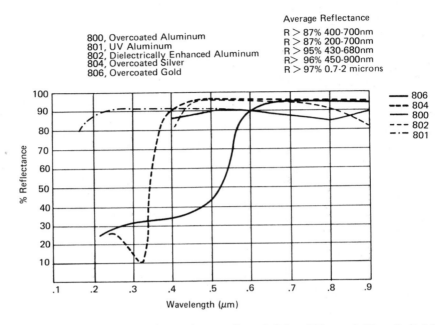

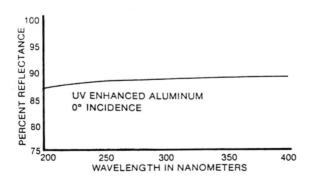

**Figure 9.5** Front surface mirrors for visible, UV, and IR. Suitable for glass, plastic, or metal substrates (from Evaporated Coatings, 1983).

**Figure 9.6** Reflectance of UV enhanced aluminum silver (from Melles Griot, 1981). Reprinted by permission of Melles Griot.

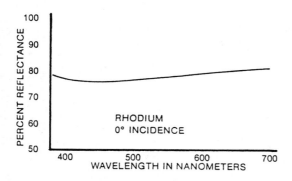

**Figure 9.7** Reflectance of rhodium (from Melles Griot, 1981). Reprinted by permission of Melles Griot.

**Table 9.3** Conventional Coatings for Mirrors

| Coating material | Comment |
|---|---|
| Al + SiO thin film | SiO coating thickness = $\lambda/2$; used in visible |
| Al + MgF$_2$ thin film | Used in UV |
| Pt, Ir, Ge, or In | Used in UV |
| Au | Used in IR |
| Rh | Resists saltwater corrosion |
| Dielectric multilayer | Used for dichroic mirrors |

**Table 9.4** Properties of Candidate Materials for Cooled Mirror Substrates[a]

| Candidate material | Relative performance parameter[b] | | | | | |
| --- | --- | --- | --- | --- | --- | --- |
| | $(k/\alpha) \times 10^{-6}$ | $E \times 10^{-6}$ | $(E/\rho) \times 10^{-6}$ | $k/\alpha E$ | $\sigma_{my} \times 10^{-3}$ | $\rho$ |
| Carbon/carbon (chopped fiber) | 300 | 2.5 | 34 | 120 | 0.25 | 0.072 |
| Silicon | 50 | 24 | 286 | 2.08 | 16 | 0.084 |
| Tungsten | 38 | 59 | 86 | 0.65 | 60 | 0.70 |
| Molybdenum | 36 | 47 | 127 | 0.55 | 35 | 0.37 |
| Copper | 25 | 17 | 53 | 1.47 | 3 | 0.32 |
| Silicon carbide | 12 | 55 | 500 | 0.22 | 30 | 0.11 |
| Beryllium | 12 | 42 | 627 | 0.29 | 17 | 0.067 |
| Aluminum | 8 | 10 | 100 | 0.80 | 25 | 0.10 |
| Be copper | 7 | 19 | 63 | 0.37 | 25 | 0.30 |
| Nickel | 5 | 30 | 94 | 0.17 | 10 | 0.32 |

[a]Source: Anthony and Hapkins (1981). F. M. Anthony and A. K. Hopkins, Actively cooled silicon mirrors, Emerging Optical Materials, Proc. SPIE 297, 196-203 (1981).

[b]High values desired except for density. $k/\alpha$, thermal growth and bowing distortion parameter; E, pressure and bowing distortion parameter; $E/\rho$, natural frequency and inertia loading parameter; $k/\alpha E$, thermal stress parameter; $\sigma_{my}$, microyield stress, estimated in most cases, permanent distortion parameter; $\rho$, density, mass parameter (where k is thermal conductivity, $\alpha$ the coefficient of thermal expansion, and E the modulus of elasticity).

## 9.4 COOLED MIRRORS

When mirrors in laser systems become excessively hot due to energy absorption, then the application of cooling becomes necessary. The various candidate substrates for cooled mirrors are listed in Table 9.4 along with a number of relative performance parameters. All of the parameters shown, except for density, are more favorable as they become larger (Anthony and Hopkins, 1981).

## REFERENCES

F. M. Anthony and A. K. Hopkins, Actively cooled silicon mirrors, Proc. SPIE *297*, Bellingham, WA, 1981.

H. Bennett, Optical properties of optical materials, Opt. Eng. *17*(5), 1978.

J. A. Dobrowolski, Coatings and filters, *Handbook of Optics*, W. G. Driscoll and W. Vaughan, Eds., McGraw-Hill Book Co., New York, 1978.

Evaporated Coatings, Inc., *Coating Data Sheet*, Coatings 800, 801, 802, 804, 806, Huntingdon Valley, PA, 1983.

Y. Namba, Specular spectral reflectance of AISI 304 stainless steel at near normal incidence, Proc. SPIE *362*, Bellingham, WA, 1982.

Melles Griot, *Optics Guide 2*, Irvine, CA, 1981.

Schott Glass Technologies, Inc., "ZERODUR Glass Ceramics," Duryea, PA, 1982.

A. Slomba, Laser mirror requirements, *Optical Industry and Systems Purchasing Directory*, Optical Publishing Co., Inc., Pittsfield, MA, 1980.

Y. S. Touloukian, R. W. Powell, C. Y. Ho, P. G. Klemens, "Thermal Properties of Matter," Vols. 1 and 2, IFI/ Plenum, NY, 1970.

# 10
# LASER MEDIA-SOLID STATE

## 10.1 THE LASER PROCESS

Luminescence is the absorption of energy in matter and its re-emission as visible or near visible radiation. If the emission occurs within $10^{-8}$ sec of the excitation, the process is called fluorescence; if greater than $10^{-8}$ sec, phosporescence. The actual delay may be microseconds to hours. Crystalline luminescent solids are known as phosphors (Kittel, 1966).

The lasing element of a solid state laser is a luminescent solid in which the light emitted in the fluorescence of one excited center stimulates other centers to emit in phase and in the same direction as the light from the first center.

A solid state laser requires a lasing medium such as a laser rod, end mirrors and a pumping source such as schematically indicated in Fig. 10.1. The mirror on one side of the rod is near 100% reflective at the wavelength of interest while the other end is partially transmittive so that some of the laser light can be emitted to perform a function. It is the light beam reflecting between these two parallel mirrors which is amplified.

This process is called light amplification by stimulated emission of radiation (LASER). The device in which this occurs is termed a "laser," and the verb "to lase" is derived from this set of ideas and terms.

Figure 10.2 illustrates a three level laser system, while Fig. 10.3 shows a four level system. The energy levels shown represent the allowable quantum energies which the ionized electrons may possess. In the three-level system the atoms are ionized by an external energy source such as a flash lamp (optical pump) to raise the electrons from the ground state E1 to an energy level of E3. From E3 the excited electrons decay to a lower level, E2. As

200

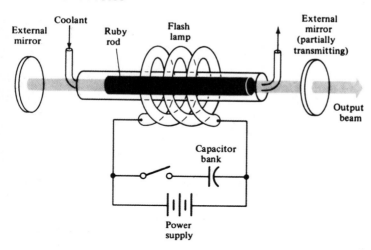

**Figure 10.1**  Typical setup of a pulsed ruby laser using flash-lamp pumping and external mirrors (from Yariv, 1971).

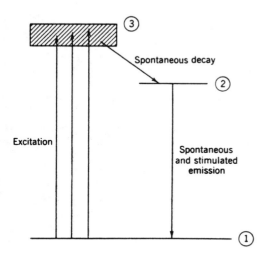

**Figure 10.2**  Simplified energy level diagram for a three-level lasing system (from Kingery, Bowen, and Uhlmann, 1976).  Reprinted with permission from W. D. Kingery, H. K. Bowen, and D. R. Uhlmann, *Introduction to Ceramics*, John Wiley and Sons, Inc., New York, 1976.

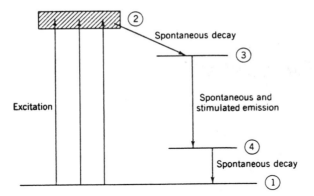

Figure 10.3 Simplified energy level diagram for a four-level lasing system (from Kingery, Bowen, and Uhlmann, 1976). Reprinted
with permission from W. D. Kingery, H. K. Bowen, and D. R. Uhlmann, *Introduction to Ceramics*, John Wiley and Sons, Inc., New
York, 1976.

the electrons spontaneously decay to the ground state, E1, photons
are emitted with wavelength equal to (E2 − E1)/h, where h is
Planck's constant.

However, if this decay is stimulated by a photon with the same
wavelength (E2 − E1)/h, then the transition takes place in phase
with the stimulating photon and in the same direction. If there
is a sufficiently large reservoir or population of electrons at the
E2 level, then a stream of photons (a light beam) passing through
the medium will increase in intensity because the gain exceeds the
absorption loss in this process. In Fig. 10.3, a similar process
occurs except the the lasing takes place between E3 and E4.

Various materials are required to satisfy the requirements of
the various optical and electro-optic elements in a laser system.
These include (i) the constituents of the laser cavity, (ii) the elements which control the laser wavelength, (iii) the devices which
define the pulse duration, and (iv) the components which control
the beam polarization.

## 10.2 LASER HOSTS

Various solid state laser hosts are used to contain the lasing ions.
The host must have near zero absorptance at the wavelength of

interest to minimize loss of useful light and to minimize thermal
input to the host from the light. In addition, high thermal conductivity of the host is desired to reduce local temperature rise,
thus minimizing hot spots and material damage. The lasing substance is contained within the structure of the host.

One of the earliest laser rods is the ruby laser composed of
single crystal $Al_2O_3$, sapphire, containing approximately 0.05%
(500 ppm) Cr. In a ruby laser, most of the energy of the flash
is dissipated as heat, but a small fraction is absorbed by the rod.
Emission takes place in a narrow band at 6943 Å (trivalent Cr ion)
which corresponds to the transition between levels 2 and 1 in Fig.
10.2. The physical properties of ruby are summarized in Table 10.1.

**Table 10.1** Properties of Ruby Laser Material[a]

| | |
|---|---|
| Density | $3.98 \ g/cm^3$ |
| Melting point | 2040°C |
| Modulus of elasticity | $50 \times 10^6$ psi |
| Modulus of rupture in bending (60-deg orientation) | $65 \times 10^3$ psi |
| Compressive strength | $300 \times 10^3$ psi |
| Hardness | 9 Mohs scale; 2000 Knoop |
| Coefficient of thermal expansion (60-deg orientation) | |
| 20-50°C | $5.8 \times 10^{-6}$ cm/cm°C |
| 20-500°C | $7.7 \times 10^{-6}$ cm/cm°C |
| Thermal conductivity (60-deg orientation) | |
| at 0°C | 0.11 cal/cm, °C, sec |
| at 100°C | 0.06 cal/cm, °C, sec |
| at 400°C | 0.03 cal/cm, °C, sec |
| Refractive index 7000 Å | 1.7638 ordinary ray |
| | 1.7556 extraordinary ray |
| Birefringence | 0.008 |
| Refractive index dopant coefficient ($\Delta n/\% \ Cr_2O_3$) | $3 \times 10^{-3}$ |
| Fluorescent lifetime, 0.05% $Cr_2O_3$ | 3 ms at 300 K |
| Fluorescent linewidth ($R_1$) | 5.0 Å at 300 K |
| Output wavelength, $R_1$ | 694.3 nm |
| Major pump bands | 404 nm (blue) and 554 nm (green) |

[a]Source: Union Carbide (1981a). Reprinted by permission of
Union Carbide Electronic Materials.

**Table 10.2**  Properties of Nd:Yttrium Aluminum Garnet[a]

| | |
|---|---|
| Chemical formula | $Y_3Al_5O_{12}$:Nd |
| Crystal structure | cubic |
| Lattice constant | 12.01 A |
| Melting point | 1970°C |
| Hardness | |
|   Mohs scale | 8.5 |
|   Vickers [111] | 1548 |
| Specific gravity | 4.56 ± .04 |
| Water absorption | zero |
| Solubility | |
|   Water | insoluble |
|   Common acids | slightly soluble |
| Thermal expansion coefficient | |
|   [100] orientation | $8.2 \times 10^{-6}$ °C$^{-1}$, 0–250°C |
|   [110] orientation | $7.7 \times 10^{-6}$ |
|   [111] orientation | $7.8 \times 10^{-6}$ |
| Thermal conductivity | |
|   20°C | 0.0320 cal sec$^{-1}$ °C$^{-1}$ cm$^{-1}$ |
|   40°C | 0.0290 |
|   100°C | 0.0250 |
|   200°C | 0.0225 |
| Specific heat capacity (0–20°C) | 0.140 cal g$^{-1}$ °C$^{-1}$ |
| Thermal shock resistance parameter (20°C)[b] | 2.6 |
| Modulus of elasticity | $45 \times 10^6$ psi |
| Tensile strength | $(20–30) \times 10^6$ psi |
| Poisson ratio | 0.3 (est.) |
| Refractive index | 1.82 ± 0.003 (d) |

[a]Source:  Union Carbide (1981b).  Reprinted by permission of Union Carbide Electronic Materials.

[b]Thermal shock resistance parameter $KS_T(1 - \mu)/E\alpha$ where K is the thermal conductivity, $S_T$ tensile stress, $\mu$ poisson ratio, E mosulus of elasticity, and $\alpha$ coefficient of thermal expansion.

A second crystalline laser widely used in neodymium in a $Y_3Al_5$-$O_{13}$ yttria garnet host (Nd—YAG), which is a four-level system (Fig. 10.3) emitting radiation at about 1.06 $\mu$m. The $Nd^{3+}$ ion has unique and favorable quantum states. YAG is a good choice for the host for continuous wave $Nd^{3+}$ lasers, while various glass hosts have been found to be more suitable for high peak power pulsed lasers.

$Y_3Al_5O_{13}$ (YAG) is a cubic crystal. $Nd_xY_{3-x}Al_5O_{22}$ (Nd:YAG) can be grown by the Verneuil method or by the Czochralski method. The optimum concentration of Nd corresponds to about 1% replacement of Y by Nd. The physical properties of YAG are summarized in Table 10.2.

Glasses are also widely used although these have low thermal conductivity and are therefore not suitable for high power continuous operation. The principal difference between glasses and single crystal hosts is the greater variation in environments surrounding the lasing ions in the glasses. In Nd-YAG the linewidth is about 10 Å, while in oxide glasses the linewidth is typically 300 Å.

A realistic specification for a glass host is that the index of refraction vary no more than $\pm 0.8 \times 10^{-6}$ across a 2.5-cm section. Indices of refraction can be varied between 1.5 and 2.0 for the host. The temperature coefficient of the index of refraction and the strain optic coefficient can be adjusted to produce thermally stable cavities (Levine, 1968).

Other ions in addition to $Nd^{3+}$ which have been made to lase in glasses are $Yb^{3+}$, $Er^{3+}$, $Ho^{3+}$, and $Tm^{3+}$.

In general, glasses and crystals are complementary. Glasses are more suitable for high energy pulsed operation because of their large size, physical parameter "tailorability," and the broadened fluorescent line. Crystals are better for continuous wave and high repetition rate pulse lasers because of narrow emission linewidths and higher thermal conductivity. Table 10.3 lists a number of solid state ceramic laser materials.

## 10.3 THE SEMICONDUCTOR JUNCTION LASER

Another type of solid state laser is the semiconductor junction laser. In this type of device the energy levels of significance are the conduction and valence bands of the solid semiconductor. A p—n junction exists at the interface between an n-type semiconductor and a p-type semiconductor. By driving a large current ($\sim 10^4$ A/cm$^2$) through a p-n junction, excited states are created and laser action can be induced. A commonly used semiconductor for this purpose

**Table 10.3** Some Solid State Laser Materials[a]

| Laser ion | Host material | Wavelength ($\mu$m) | Remarks |
|---|---|---|---|
| $Cr^{3+}$ | $Al_2O_3$ | 0.6943 | Operates at room temp. high power |
| $Pr^{3+}$ | $CaWO_4$ | 1.04 | Liquid air temp. |
|  | $LaF_3$ | 0.5985 |  |
| $Nd^{3+}$ | $BaF_2$ | 1.060 |  |
|  | $SrF_2$ | 1.043 | Liquid air temp. |
|  | $CaF_2$ | 1.046 |  |
|  | $CaMoO_4$ | 1.067 |  |
|  | $CaWO_4$ | 1.06 |  |
|  | $Gd_2O_3$ | 1.079 | Room temp. |
|  | $LaF_3$ | 1.063 |  |
|  | Glass | 1.06 |  |
| $Sm^{2+}$ | $CaF_2$ | 0.7085 | Liquid helium temp. |
|  | $SrF_2$ | 0.6969 |  |
| $Eu^{3+}$ | $Y_2O_3$ | 0.6113 | Room temp. |
| $Gd^{3+}$ | Glass | 0.3125 | Liquid air temp. |
| $Ho^{3+}$ | $CaF_2$ | 2.09 |  |
|  | $CaWO_4$ | 2.05 | Liquid air temp. |
|  | Glass | 2.04 |  |
| $Er^{3+}$ | $CaF_2$ | 1.617 | Liquid air temp. |
|  | $CaWO_4$ | 1.612 |  |
| $Tm^{3+}$ | $SrF_2$ | 1.972 | Liquid air temp. |
|  | $CaWO_4$ | 1.911 |  |
| $Yb^{3+}$ | Glass | 1.015 | Liquid air temp. |
| $U^{3+}$ | $BaF_2$ | 2.55 | Liquid air temp. |
|  | $CaF_2$ | 2.51-2.61 | Room temp. |
|  | $SrF_2$ | 2.40 | Liquid air temp. |

[a]Source: Fowles (1968). From *Introduction to Modern Optics* by G. R. Fowles, Hold, Rinehart and Winston, Inc., 1968. Reprinted by permission of Holt, Rinehart and Winston, CBS College Publishing.

is GaAs which lases at 8370 Å at 4.2 K. A schematic sketch of such a device is shown in Fig. 10.4, while Table 10.4 lists semi-conductors which can exhibit lasing action.

As indicated in Fig. 10.4 the lasing region is a narrow slice of material at the junction on the order of 2 μm thick, which is called the active layer. This type of laser is widely used for communication through fiberoptic cable.

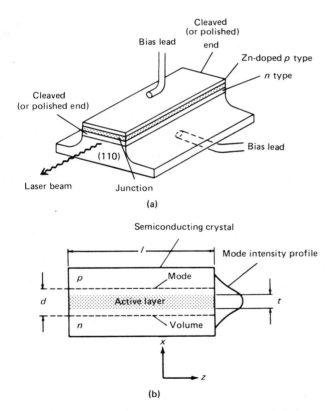

Figure 10.4 (a) Typical p–n junction laser made of GaAs. Two parallel (110) faces are cleaved and serve as reflectors. (b) schematic diagram showing active layer and transverse (x) distribution of the laser mode. (From Yariv, 1971.)

**Table 10.4** Oscillation Wavelength (Temperature)
of Some Semiconductor p–n Junction Lasers [a]

| Material | Oscillation wavelength ($\mu$m) | | |
|---|---|---|---|
| GaAs | 0.837 (4.2 K) | 0.843 (77 K) | |
| InP | | 0.907 (77 K) | |
| InAs | | 3.1 (77 K) | |
| InSb | 5.26 (10 K) | | |
| PbSe | 8.5 (4.2 K) | | |
| PbTe | 6.5 (12 K) | | |
| $Ga(As_xP_{1-x})$ | 0.65-0.84 | | |
| $(Ga_xIn_{1-x})As$ | 0.84-3.5 | | |
| $In(As_xP_{1-x})$ | 0.91-3.5 | | |
| GaSb | | 1.6 (77 K) | |
| $Pb_{1-x}Sn_xTe$ | 9.5-28 ($\sim$12 K)<br>↓ ↓<br>$x = 0.15$  $x = 0.27$ | | |

[a]Source: Yariv (1971).

In general, the lower the threshold current for lasing, the longer the life. Although continuous wave (low) operation is desirable, pulsed operation is often used because of the inability of the lasing medium to dissipate heat fast enough in the continuous mode to avoid damage to the laser medium.

## 10.4   LASER WINDOW MATERIALS

In gas lasers, where a gas is the lasing medium, the gas is confined in a tube with a window at the emitting end. The window must have very low absorptance, and be sufficiently strong and

**Table 10.5**  Typical Gas Laser Window Materials

| Laser wavelength ($\mu$m) | Material |
|---|---|
| 1-2 | $Al_2O_3$ and other oxides |
| 5 | $CaF_2$ and other alkaline halides |
| 10 | NaCl and other halides |
| 10 | ZnSe, CdTe, and other II-VI compounds |

stiff to withstand the stresses induced by the thermal gradient developed and by the differential pressure with very little distortion. Typical window materials are listed in Table 10.5.

## REFERENCES

G. R. Fowles, *Introduction to Modern Optics*, Holt, Rinehart, and Winston Inc., New York, 1968.

W. D. Kingery, H. K. Bowen, and D. R. Uhlmann, *Introduction to Ceramics*, John Wiley and Sons, New York, 1976.

C. Kittel, *Introduction to Solid State Physics*, John Wiley and Sons, Inc., New York, 1966.

A. K. Levine, *Lasers*, Marcel Dekker, Inc., New York, 1968, Vol. 2.

Union Carbide Electronics Division, *High Performance Cz Ruby Laser Rods*, San Diego, CA, 1981a.

Union Carbide Electronics Division, *High Quality Nd:YAG Laser Rods*, San Diego, CA, 1981b.

A. Yariv, *Introduction to Optical Electronics*, Holt, Rinehart and Winston, Inc., New York, 1971.

# 11

## ELECTRO-OPTIC, ACOUSTO-OPTIC, AND LIQUID CRYSTALS

### 11.1 POLARIZED LIGHT

Electromagnetic waves are transverse to the direction of propagation. Light has an electric vector and an orthogonal magnetic vector associated with the electric vector; both vectors are transverse to the axis of propagation. This concept is illustrated in Fig. 11.1. If the electric vector oscillates in one plane only, the light is linearly polarized. If linearly polarized light, in addition, contains unpolarized light, the light is then said to be partially linearly polarized.

If the electric vector is constant in magnitude and its tip describes a circular motion about the axis of propagation, the light is circularly polarized. Elliptical polarization stands between linear and circular polarization.

In a birefringent crystal, light has different velocities depending on the plane of vibration of the electric vector relative to the axis of the crystal. The classic example of birefringence, discovered by Erasmus Bartholinus in 1669, is calcite, $CaCO_3$. If unpolarized light passes through a calcite crystal as shown in Fig. 11.2 the light separates into two rays, the ordinary (o) ray and the extraordinary (e) ray. The o ray obeys Snell's law of refraction but the e ray does not. The two rays are each linearly polarized at right angles to each other.

### 11.2 THE ELECTRO-OPTIC EFFECT

The electro-optic effect (Pockels effect) is the change in the indices of refraction of the ordinary and the extraordinary rays that is caused by and is proportional to an applied electric field. The electro-optic effect was discovered by Roentgen and extensively

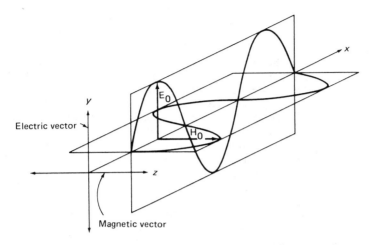

**Figure 11.1** Electromagnetic field represented by electric and magnetic vectors (from Meyer-Arendt, 1972). R. Meyer-Arendt, *Introduction to Classical and Modern Optics*, Prentice-Hall, Inc., Englewood Cliffs, N.J., 1972, pp. 281 and 304. Reprinted by permission of Prentice-Hall, Inc., Englewood Cliffs, N.J.

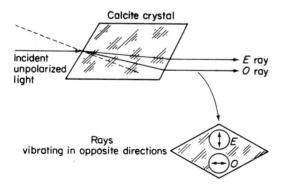

**Figure 11.2** Double refraction and polarization of light in calcite (from Meyer-Arendt, 1972). R. Meyer-Arendt, *Introduction to Classical and Modern Optics*, Prentice-Hall, Inc., Englewood Cliffs, N.J., 1972, pp. 281 and 304. Reprinted by permission of Prentice-Hall, Inc., Englewood Cliffs, N.J.

investigated by F. Pockels around the turn of the nineteenth century. This effect occurs in certain crystals known as electro-optic crystals.

The electro-optic effect provides a convenient means of controlling intensity or phase of the propagating radiation. The effect is characterized by a tensor with coefficients $r_{ij}$, where $i = 1, 2, 3$ and $j = 1, \ldots, 6$. There are 18 such coefficients but depending on the symmetry of the crystal many of the $r_{ij}$ terms drop out. These coefficients can be measured; they provide information necessary to compute the electrically induced birefringence. Some of the materials which exhibit the electro-optic effect and are used in electro-optic devices are listed in Table 11.1.

A KDP longitudinal modulator is illustrated in Fig. 11.3. When no voltage is applied, the indices of refraction for the o and e rays are identical along the optic axis. The incoming beam is plane-polarized with the plane of polarization aligned to the x or y axis. When a voltage is applied parallel to the light beam, the crystal becomes birefringent.

Table 11.1  Characteristics of Longitudinal Modulators[a]

| Material | Electro-optic constant $r_{63}$ (pm/V) | Typical half-wave voltage at 546 nm (kV) | Approximate $n_o$ |
|---|---|---|---|
| Ammonium dihydrogen phosphate (ADP) | 8.5 | 9.2 | 1.526 |
| Potassium dihydrogen phosphate (KDP) | 10.5 | 7.5 | 1.51 |
| Potassium dideuterium phosphate (KD*P) | 26.4 | 2.6-3.4 | 1.52 |
| Potassium dihydrogen arsenate (KDA) | 10.9 | 6.4 | 1.57 |
| Rubidium dihydrogen phosphate (RDP) | 11.0 | 7.3 | |
| Ammonium dihydrogen arsenate (ADA) | 5.5 | 13 | 1.58 |

[a]Source:  Hartfield and Thompson (1978).  Reprinted with permission of W. G. Driscoll and W. Vaughan, *Handbook of Optics*, McGraw-Hill Book Company, New York, 1978.

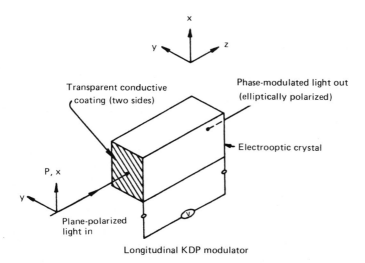

Longitudinal KDP modulator

Figure 11.3 Longitudinal KDP Modular (from Hartfield and Thompson, 1978). Reprinted with permission of W. G. Driscoll and W. Vaughan, *Handbook of Optics*, McGraw-Hill Book Company, New York, 1978.

The extraordinary ray is retarded with respect to the ordinary ray, and upon emerging from the output face of the electro-optic crystal, the two polarized beams have acquired a relative phase shift due to the retardation of the light in the plane of the slow axis, resulting in an elliptically polarized beam.

The retardation is given by

$$\Phi = \frac{n_o^3 r_{63} V}{\lambda}$$

where $\Phi$ is the number of wavelengths retardation, $n_o$ the ordinary index of refraction, $r_{63}$ the electro-optic coefficient ($\mu m/V$), v the applied voltage (V), and $\lambda$ the wavelength of incident light ($\mu m$).

For the value $\Phi = 1/2$, $V = V_{1/2}$ is defined as the half-wave voltage. Table 11.1 provides half-wave voltages for a number of electro-optic crystals. The half-wave voltage is a function of the wavelength, as is illustrated in Fig. 11.4 for a KD*P crystal.

Figure 11.5 illustrates a simple arrangement for modulating a light beam using an electro-optic modulator between crossed polarizers. The output intensity can be made to linearly replicate a

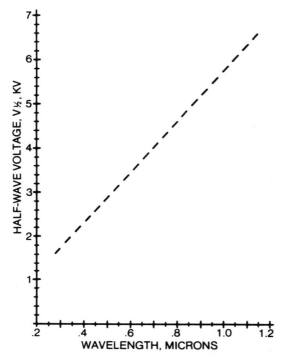

Figure 11.4 Half-wave voltage as a function of wavelength for KD*P (from Interactive Radiation, Inc.).

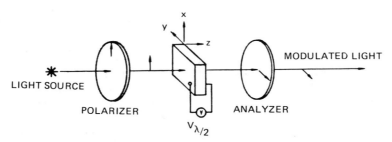

Figure 11.5 Simple set-up for intensity modulating a light beam using an electro-optic modulator between crossed polarizers (from Goldstein, 1968).

**Table 11.2**  Characteristics of KDP, KD*P, and ADP[a]

| | $KH_2PO_4$ (KDP) | $KD_2PO_4$ (KD*P) | $NH_4H_2PO_4$ (ADP) |
|---|---|---|---|
| **Electro-optic constant at** | | | |
| **546 nm (pm/V)** | | | |
| $r_{41}^T$ | 8.77 | 8.8 | 24.5 |
| $r_{41}^S$ | — | — | — |
| $r_{63}^T$ | 10.5 | 26.4 | 8.5 |
| $r_{63}^S$ | 9.7 | — | 5.5 |
| **Dielectric constant** | | | |
| $\varepsilon_{33}^T/\varepsilon_0$ | 21 | 50 | 15.4 |
| $\varepsilon_{33}^S/\varepsilon_0$ | 21 | 48 | 13.8 |
| $\varepsilon_{11}^T/\varepsilon_0$ | 42 | — | 56 |
| $\varepsilon_{11}^S/\varepsilon_0$ | 42 | — | 56 |
| **Loss tangent, constant** | | | |
| **strain at 9.2 GHz** | | | |
| $\tan\delta_3$ | 0.0075 | 0.11 | 0.006 |
| $\tan\delta_1$ | 0.0045 | 0.025 | 0.007 |
| **Curie temperature (K)** | 122 | 221 | 148 |
| **Piezoelectric constant (pm/V)** | | | |
| $d_{14}$ | +1.3 | — | -1.5 |
| $d_{36}$ | +21 | 58 | +48 |
| **Elasto-optic constant** | | | |
| $p_{66}$ | -0.068 | — | -0.075 |
| $p_{44}$ | — | — | -0.051 |
| **Maximum safe operating** | | | |
| **temperature (°C)** | 100 | 100 | 80 |

[a] Source:  Hartfield and Thompson (1978).  Reprinted with permission of W. G. Driscoll and W. Vaughan, *Handbook of Optics*, McGraw-Hill Book Company, New York, 1978.

Table 11.3 Properties of Electro-Optic Materials at Room Temperature[a]

| Crystal structure and material | Optical symmetry | Refractive | |
|---|---|---|---|
| | | $n_o$ | $n_e$ |
| Tetragonal | | | |
| ammonium dihydrogen phosphate (ADP), $NH_4H_2PO_4$ | Uniaxial | 1.526 | 1.481 |
| potassium dihydrogen phosphate (KDP), $KH_2PO_4$ | Uniaxial | 1.512 | 1.470 |
| potassium dideuterium phosphate (KD*P or DKDP), $KD_2PO_4$ | Uniaxial | 1.508 | 1.468 |
| potassium dihydrogen arsenate (KDA), $KH_2AsO_4$ | Uniaxial | 1.571 | 1.521 |
| ammonium dihydrogen arsenate (ADA), $NH_4H_2AsO_4$ | Uniaxial | 1.578 | 1.522 |
| rubidium dihydrogen arsenate $RbH_2AsO_4$ | Uniaxial | 1.559 | 1.520 |
| rubidium dihydrogen phosphate $RbH_2PO_4$ | Uniaxial | | |
| barium titanate, $BaTiO_3$ | Uniaxial | 2.437 | 2.365 |
| Cubic perovskite | | | |
| barium titanate, $BaTiO_3$ | Isotropic | 2.43 | |
| strontium titanate, $SrTiO_3$ | Isotropic | 2.38 | |
| potassium tantalate, $KTaO_3$ | Isotropic | 2.24 | |
| potassium tantalate niobate (KTN), $KTa_{0.65}Nb_{0.35}O_3$ | Isotropic | 2.29 | |
| lead magnesium niobate $Pb_3MgNb_2O_9$ | Isotropic | 2.56 | |
| Rhombohedral perovskite | | | |
| lithium niobate (LN), $liNbO_3$ | Uniaxial | 2.297 | 2.208 |
| lithium tantalate (LT), $LiTaO_3$ | Uniaxial | 2.183 | 2.188 |

| index | Half-wave voltage[b] | | | $\lambda_1$ to $\lambda_2$,[d] | Transmission region |
| Wavelength $\lambda$ (μm) | $V_{\lambda/2}$ (kV) | Wavelength $\lambda$ (μm) | Direction[c] | (μm) | (μm) |
|---|---|---|---|---|---|
| 0.546 | 9.0<br>7.65 | 0.546 | L<br>T | 0.4-0.7 | 0.125-1.7 |
| 0.546 | 7.65<br>17.8 | 0.546 | L<br>T | 0.436,0.578 | 0.250-1.7 |
| 0.546 | 2.98<br>8.00 | 0.546 | L<br>T | 0.5-0.75 | 0.19-2.15 |
| 0.546 | 6.43 | 0.546 | L | 0.436,0.578 | 0.246->0.75 |
| 0.589 | 13.0 | 0.546 | L | | 0.260->0.75 |
| 0.546 | 5.54 | 0.546 | L | 0.436,0.578 | |
| | 7.3 | 0.546 | L | | |
| 0.546 | 0.38 | 0.546 | T | | |
| Visible | 0.66[e]<br>$T > 120°C$ | 0.546 | T | 0.4-1.0<br>$(g_{11}-g_{12})$ | ~0.4-7.5 |
| 0.633 | ~13[e] | | T | | 0.39-6.8 |
| 0.633 | 0.95 at 4.2 K[e] | 0.633 | T | | |
| 0.633 | 0.58[e] | 0.633 | T | 0.4-2.0<br>$(g_{11}-g_{12})$ | 0.39-6 |
| 0.633 | | | T | | 0.5-5 |
| 0.60 | 2.940<br>2.5 | 0.633 | T<br>L | | 0.4-5 |
| 0.60 | 2.800 | 0.633 | T | | 0.45-5 est |

Table 11.3 (Continued)

| Crystal structure and material | Optical symmetry | Refractive | |
|---|---|---|---|
| | | $n_o$ | $n_e$ |
| Monoclinic | | | |
| calcium pyroniobate, $CaNb_2O_7$ | Biaxial | 1.97, 2.16, | 2.17 |
| Orthorhombic tungsten bronze | | | |
| barium sodium niobate, | Uniaxial | 2.326, 2.324, | 2.221 |
| $Ba_2NaNb_5O_{15}$ | | | |
| Tetragonal tungsten bronze | | | |
| potassium lithium niobate, | Uniaxial | 2.28 | 2.13 |
| $(K_2O)_{0.3}(Li_2O)_{0.7-x}(Nb_2O_5)_x$ | | $x = 0.52$ | |
| $0.515 \leqslant x \leqslant 0.55$ | | | |
| potassium lithium niobate, | Uniaxial | 2.277 | 2.163 |
| $K_{0.6}Li_{0.4}NbO_3$ or $K_3Li_2Nb_5O_{15}$ | | | |
| potassium sodium barium niobate | Uniaxial | 2.315 | 2.230 |
| $K_xNa_{1-x}Ba2Nb5O15$, | | | |
| $0.70 \leqslant x \leqslant 0.90$ | | | |
| potassium strontium niobate | Uniaxial | ~2.25 | ~2.25 |
| $KSr_2Nb_5O_{15}$ | | | |
| strontium barium niobate | Uniaxial | 2.312 | 2.299 |
| $Sr_xBa_{1-x}Nb_2O_6$, $0.25 \leqslant x \leqslant 0.75$ | | | |
| Cubic zinc blende | | | |
| Cadmium telluride, CdTe | Isotropic | 2.6 | |
| cuprous chloride, CuCl | Isotropic | 1.996 | |
| cuprous bromide, CuBr | Isotropic | 2.16 | |
| gallium arsenide, GaAs | Isotropic | 3.42 | |
| gallium phosphide, GaP | Isotropic | 3.350 | |
| zinc sulfide, ZnS | Isotropic | 2.364 | |

| index | Half-wave voltage[b] | | | | |
|---|---|---|---|---|---|
| Wave-length $\lambda$ ($\mu$m) | $V_{\lambda/2}$ (kV) | Wave-length $\lambda$ ($\mu$m) | Direction[c] | $\lambda_1$ to $\lambda_2$,[d] ($\mu$m) | Transmission region ($\mu$m) |
| | 4.55 | 0.633 | T | | |
| 0.633 | 1.570 | 0.633 | T | | 0.4-5 |
| 0.633 | 1.350 | 0.633 $0.525 \leqslant x \leqslant 0.55$ | T | | |
| 0.633 | 0.930 | 0.633 | T | | 0.4-5 |
| 0.633 | 1.410 | 0.633 $x = 0.8$ | T | | |
| 0.633 | 0.43 | 0.633 | T | | |
| 0.633 | 0.037 | 0.633 $x = 0.75$ | T | | |
| 1-30 | 4.9 | 1.0 | T | 3.39, 10.6, 23.35, 27.95 $(n_o{}^3 r_{41})$ | 1-30 |
| 0.535 | 6.2 6.2 | 0.546 | L T | | 0.4-20 |
| 0.535 | | | T | 0.525, 0.675 $(n_o{}^3 r_{41})$ | |
| 1.0 | 4.5 | 1.0-1.7 | T | 10.6 $(r_{41})$ | 1.0-15 |
| 0.60 | 13 | 0.546 | T | | 0.6-4.5 |
| 0.60 | 10.2 | 0.546 | T | 0.404-0.644 $(r_{41})$ | |

Table 11.3 (Continued)

| Crystal structure and material | Optical symmetry | Refractive $n_o$ | $n_e$ |
|---|---|---|---|
| Cubic zinc blende (continued) | | | |
| zinc selenide, ZnSe | Isotropic | 2.654 | |
| zinc telluride, ZnTe | Isotropic | 3.06 | |
| | | | |
| Cubic euyltine | | | |
| bismuth germanate, $Bi_4(GeO_4)_3$ | Isotropic | 2.07 | |
| Body-centered | | | |
| bismuth germanium oxide, $Bi_{12}GeO_{20}$ | Isotropic | 2.55 | |
| Tetragonal | | | |
| cadmium mercury thiocyanate, $Cd[Hg(SCN)_4]$ | Uniaxial | | |
| Hexagonal wurtzite | | | |
| cadmium sulfide, CdS | Uniaxial | 2.493 | 2.511 |
| Hexagonal | | | |
| crystalline quartz, $SiO_2$ | Uniaxial | 1.544 | 1.553 |
| Cubic | | | |
| hexamethylenetetramine (HMT or hexamine), $N_4(CH_2)_6$ | Isotropic | 1.591 | |

[a]*Source*: Bennett and Bennett (1978).
[b]When values of $V_{\lambda/2}$ are given for more than one direction in the crystal, the minimum value is reported here.
[c]L = longitudinal; T = transverse.
[d]Other wavelengths at which half-wave voltage has been measured.
[e]Quadratic effect; $V_{\lambda/2}$ for first $\pi$ rad retardation in a 1-cm cube.

| index | Half-wave voltage[b] | | | $\lambda_1$ to $\lambda_2$,[d] | Transmission region |
| Wave-length $\lambda$ ($\mu$m) | $V_{\lambda/2}$ (kV) | Wave-length $\lambda$ ($\mu$m) | Direction[b] | ($\mu$m) | ($\mu$m) |
|---|---|---|---|---|---|
| 0.546 | 7.1 | 0.5475 | T | | |
| 0.589 | 2.3 | 0.590 | T | 0.59-0.69<br>$10.6(r_{41})$ | 0.57-52 |
| | 14.8 | 0.543 | T | 0.453,0.616 | 0.35-6 |
| 0.51 | 12.0 | 0.666 | T | | |
| | 15.6 | 0.633 | T | 0.476 | |
| 0.60 | 8.6 | 0.589 | L | | 0.5-16 |
| 0.589 | 325 | 0.546 | T | | 0.12-4.5 |
| 0.589 | 8.3 | 0.5475 | T | | 0.35-2.25 |

modulating voltage. To do this, one half of $V_{1/2}$, $V_{1/4}$, is applied
to the crystal. This causes a quarter wave retardation of the vertically polarized light entering the electro-optic crystal. If a small
modulating ac voltage is now superimposed on the $V_{1/4}$ voltage,
then the modulating voltage results in a variation of angle of polarization linear with respect to the variations in the modulating voltage. In the arrangement of polarizer and analyzer shown in Fig.
11.5, the output light intensity reproduces the variations in modulation voltage.

A transverse electro-optic effect is also exhibited by some crystals. In this case the birefringence changes are induced by a voltage applied transversely to the beam propagation direction as
indicated in Fig. 11.6. The electric field E is applied parallel to
the z axis. The input beam plane of polarization is 45° to the z axis,
and the crystal length is l. The half-wave voltage is given by

$$V_{\lambda/2} = \frac{(El)_{\lambda/2}}{l/a}$$

$(El)_{\lambda/2}$ is the half-wave field-distance product and is defined for
a given cube of material. For LiTaO$_3$, $(El)_{\lambda/2} \sim 2700$ V at a wavelength $\lambda = 633$ nm.

Many solid and liquid materials exhibit the electro-optic effect.
The problem of growing more uniform and larger electro-optic

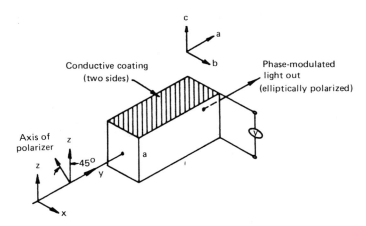

Figure 11.6 Transverse LiTaO$_3$ modulator (from Hartfield and
Thompson, 1978). Reprinted with permission of W. G. Driscoll
and W. Vaughan, *Handbook of Optics*, McGraw-Hill Book Company,
New York, 1978.

crystals is the subject of current research. Problems associated with crystal quality are avoidance of variations of refractive index due to residual stress, nonstoichiometry, crystal growth faults, and cumulative damage by the laser beam.

Some characteristics of KDP, KD*P, and ADP are shown in Table 11.2, and an extensive list of electro-optic crystals and some of their properties are shown in Table 11.3.

ADP, KDP, and KD*P can be cut and polished but are water soluble and are easily damaged. The KD*P (where deuterium is substituted for hydrogen in the KDP crystal) is superior to the KDP because only about one third the voltage is required to achieve a $\lambda/2$ retardation in the longitudinal mode.

Materials of prime importance are ADP, KDP, KD*P, KTN, $LiNbO_3$, $LiTaO_2$, $Ba_3NaNb_5O_{15}$, $Sr_xBa_{1-x}Nb_2O_6$, CuCl, ZnSe, hexamethylenetetramine, and Se.

Some materials have a quadratic rather than linear electro-optic effect. That is, the retardation is proportional to the square of the applied voltage. Among these are strontium titanate, potassium tantalate, and KTN, all at room temperature. These crystals are all of cubic structure. The $\lambda/2$ voltages, $V_{\lambda/2}$, given in Table 11.3 are for the first $\pi$ radian retardation for quadratic crystals.

All crystals which are linearly electro-optic are also piezoelectric. That is, the application of an electric field will produce a change in dimension, or, if pressure is applied, a voltage will be produced. The application of an electric field thus produces a phase shift in addition to the electro-optical phase shift due to this change in the dimensions of the crystal.


## 11.3  PIEZO-OPTIC EFFECT

The Piezo-optic effect is analagous to the electro-optic effect. When pressure is applied to a piezo-optic crystal, the birefringence of the materials changes. Some of these piezo-optic crystals are isotropic and cubic, but when stressed become birefringent. If such a crystal is stressed in a direction perpendicular to a light beam, the phase retardation is proportional to the stress. This effect is applied in acousto-optic modulators.


## 11.4  ACOUSTO-OPTIC MODULATORS

Acousto-optic devices are based on the fact that refraction and diffraction effects occur when light passes through a solid medium transversely to a high frequency (ultrasonic) acoustic field propagating in the same medium. The interaction can affect beam deflection, polarization, phase, frequency or amplitude of the optical beam. Ultrasonic waves range from the $10^5$ Hz (above the upper range, approximately $2 \times 10^4$ Hz, of the human ear) to $10^9$ Hz.

The acoustic wave, because of the piezo-optic effect, induces a cyclic change in the index of refraction. Light passing through such a crystal in a direction transverse to the acoustic wave direction is modulated by the wave. For example, in a fused quartz element, 10 cm thick, oscillating at 50 KHz, a transverse light beam at 4000 Å will experience a phase retardation of $\pi$ radians. The acoustic wave is induced by a peizoelectric driver. A schematic illustration of diffraction of light by an acoustic wave is given in Fig. 11.7.

Table 11.4 lists some of the more important acousto-optic materials and some of their characteristics. There are two figures of merit used in comparing these materials, namely,

$$F_1 = \frac{n^6 p^2}{\rho v^3}$$

$$F_2 = \frac{n^7 p^2}{\rho v}$$

where n is the optical index of refraction, p is the appropriate component of the photoelastic tensor (analogous to r in the electro-

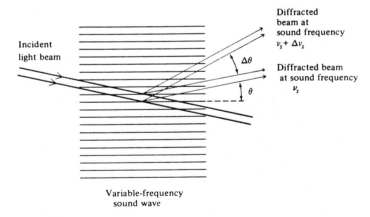

Figure 11.7 Diffraction of a light beam by an acoustic wave—a change in frequency of the sound wave from $v_s$ to $\Delta v_s$ causes a change $\Delta \theta$ in the direction of the diffracted beam (from Yariv, 1971).

Table 11.4 Figures of Merit for Acousto-Optic Crystals[a]

| Material | $\lambda$, μm | $n$ | Polarization and direction[b] | | Figure of merit | |
| --- | --- | --- | --- | --- | --- | --- |
| | | | Acoustic wave | Optical wave[c] | $\dfrac{n^6 p^2}{\rho v^3} \times 10^{-18}$ | $\dfrac{n^7 p^2}{\rho v} \times 10^{-7}$ |
| Fused quartz | 0.63 | 1.46 | L | ⊥ | 1.51 | 7.89 |
| | | | T | ∥ or ⊥ | 0.467 | 0.963 |
| GaP | 0.63 | 3.31 | L, [110] | ∥ or ⊥, [010] | 44.6 | 590 |
| | | | T, [110] | ∥ or ⊥, [010] | 24.1 | 137 |
| GaAs | 1.15 | 3.37 | L, [110] | ∥ or ⊥, [010] | 104 | 925 |
| | | | T, [100] | ∥ or ⊥, [010] | 46.3 | 155 |
| TiO$_2$ | 0.63 | 2.58 | L, [11$\bar{2}$0] | ⊥, [001] | 3.93 | 62.5 |
| LiNbO$_3$ | 0.63 | 2.20 | L, [11$\bar{2}$0] | | 6.99 | 66.5 |
| YAG | 0.63 | 1.83 | L, [100] | ∥ | 0.012 | 0.16 |
| | | | L, [110] | ⊥ | 0.073 | 0.98 |
| YIG | 1.15 | 2.22 | L, [100] | ∥ | 0.33 | 3.94 |
| LiTaO$_3$ | 0.63 | 2.18 | L, [001] | ⊥ | 1.37 | 11.4 |
| As$_2$S$_3$ | 0.63 | 2.61 | L | ⊥ | 433 | 762 |
| | 1.15 | 2.46 | L | ∥ | 347 | 619 |
| SF-4 | 0.63 | 1.616 | L | ⊥ | 4.51 | 1.83 |
| β-ZnS | 0.63 | 2.53 | L, [110] | ∥, [001] | 3.41 | 24.3 |
| | | | T, [110] | ∥ or ⊥, [001] | 0.57 | 10.6 |

**Table 11.4** (Continued)

| Material | $\lambda$, μm | $n$ | Polarization and direction[b] | | Figure of merit | |
|---|---|---|---|---|---|---|
| | | | Acoustic wave | Optical wave | $\dfrac{n^6 p^2}{p\nu^3} \times 10^{-18}$ | $\dfrac{n^7 p^2}{p\nu} \times 10^{-7}$ |
| $\alpha\text{-Al}_2\text{O}_3$ | 0.63 | 1.76 | L, [001] | ∥, [11̄20] | 0.34 | 7.32 |
| CdS | 0.63 | 2.44 | L, [11̄20] | ∥ | 12.1 | 51.8 |
| ADP | 0.63 | 1.58 | L, [100] | ∥, [010] | 2.78 | 16.0 |
| | | | T, [100] | ∥ or ⊥.[001] | 6.43 | 3.34 |
| KDP | 0.63 | 1.51 | L, [100] | ∥, [010] | 1.91 | 8.72 |
| | | | T, [100] | ∥ or ⊥,[001] | 3.83 | 1.57 |
| $\text{H}_2\text{O}$ | 0.63 | 1.33 | L | | 160 | 43.6 |
| Te | 10.6 | 4.8 | L, [11̄20] | ∥, [0001] | 4400 | 10200 |
| $\alpha\text{-HIO}_3$[d] | 0.63 | | L-a | a–c | 48.2 | |
| | | | | b–c | 20.8 | |
| | | | | c–b | 46.0 | |
| | | | L-b | a–c | 41.6 | |
| | | | | b–c | 58.9 | |
| | | | | c–a | 32.8 | |
| | | | L-c | a–b | 83.5 | |
| | | | | b–a | 77.5 | |
| | | | | c–a | 63.0 | |
| | | | Shear a–b | a–c | 17.1 | |

[a]Source: Hartfield and Thompson (1978). Reprinted with permission of W. G. Driscoll and W. Vaughan, *Handbook of Optics*, McGraw-Hill Book Company, New York, 1978.
[b]$L$ = longitudinal, $T$ = transverse (shear).
[c]Polarization is defined parallel ($\parallel$) or perpendicular ($\perp$) to the plane formed by the acoustic and optic propagation directions (k vectors).
[d]Lattice constants:  $A$ = 5.888 Å, $b$ = 7.733 Å, $c$ = 5.538 Å.

optic effect), $\rho$ is the mass density, and $\nu$ is the acoustic phase velocity.

Since $\rho, \nu$, and p exhibit no extreme variations in the various acousto-optic materials, the candidates of most interest are those with good optical and acoustic qualities with large n, although optical and acoustic parameters are not included in the figures or merit.

In Table 11.4, the longitudinal (L) wave is one where the displacement u is parallel ($\parallel$) to the acoustic wave propagation direction (k vector), whereas the transverse (T) wave is one where the displacement u is normal ($\perp$) to the acoustic wave propagation direction (k vector). The polarization of the optical wave is defined as parallel ($\parallel$) or perpendicular ($\perp$) to the plane formed by the acoustic and optic k vectors.

Alpha-iodic acid ($\alpha$-HIO$_3$) cited in Table 11.4 is reported to have, in the blue-green portion of the visible spectrum, an extremely high figure of merit, $n^6 p^2 / \rho \nu^3$, approximately twelve times that of LiNbO$_3$ and fifty times that of fused silica. Large solution-grown crystals of high quality have been made. The material has low acoustic and optic losses and is minimally damaged by laser light in the visible spectrum (Pinnow and Dixon, 1968).

Newly identified materials for acousto-optic application include alpha-aluminum phosphate ($\alpha$-AlPO$_4$, berlinite) and gallium arsenide (Ballato and Lukaszek, 1980).

## 11.5 LIQUID CRYSTALS

A spatial optical modulator, by definition, is a device that modifies the intensity of an incident light beam, two dimensionally. Such devices use electro-optic, acousto-optic or other effects. Liquid crystals are often employed for display of the output of the modulator, as well as participating in the modulation.

During the melting process, in certain materials, the three-dimensional order becomes a two-dimensional (planar) or one-dimensional (linear) state of order. Layers or strands are formed and persist over a temperature range. These materials are termed liquid crystals. Some compounds which exhibit liquid crystal characteristics are listed in Fig. 11.8.

Certain of these liquid crystals can form a nematic phase. In the nematic phase there are no interactive forces between the ends of the elongated molecules and strand-shaped molecular aggregates are formed as indicated in Fig. 11.9. Electrical and magnetic fields can alter the optical properties of neumatic liquid crystals and thus can be employed to create optical displays.

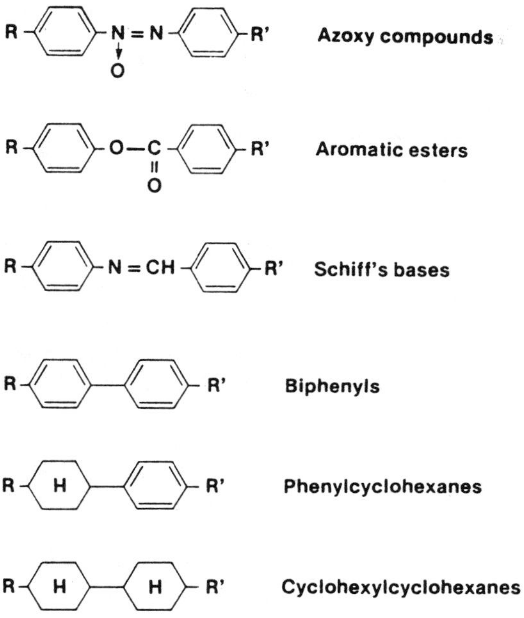

R⟨◯⟩-N=N-⟨◯⟩-R'     **Azoxy compounds**
         ↓
         O

R⟨◯⟩-O-C-⟨◯⟩-R'     **Aromatic esters**
        ‖
        O

R⟨◯⟩-N=CH-⟨◯⟩-R'     **Schiff's bases**

R⟨◯⟩-⟨◯⟩-R'     **Biphenyls**

R⟨H⟩-⟨◯⟩-R'     **Phenylcyclohexanes**

R⟨H⟩-⟨H⟩-R'     **Cyclohexylcyclohexanes**

Figure 11.8 Several compounds that can exhibit liquid crystalline properties (from Macklin, 1980). Reprinted from *The Optical Industry and Systems Purchasing Directory*, 1980.

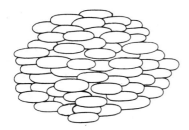

Figure 11.9 A structural diagram of a nematic phase (from Macklin, 1980). Reprinted from *The Optical Industry and Systems Purchasing Directory*, 1980.

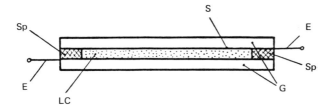

Figure 11.10 Cross section of a liquid crystal display: G represents the glass plates; SP is the spacer, LC the liquid crystal; E an electrode lead; S represents the electrode, generally $SnO_2$ or $InO_3$ (from Macklin, 1980). Reprinted from *The Optical Industry and Systems Purchasing Directory*, 1980.

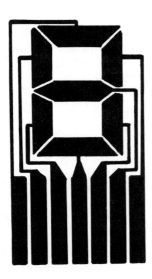

Figure 11.11 Layout of a seven-segment numerical display (from Macklin, 1980). Reprinted from *The Optical Industry and Systems Purchasing Directory*, 1980.

In a nematic display, two glass plates, coated with a pattern of transparent conducting material, are mounted with the patterns opposing each other and held about 10-20 μm apart by means of spacers. The nematic phase is placed between the glass plates and the assembly is sealed. Such an assembly is indicated in Figs. 11.10 and 11.11.

The elongated molecules can be oriented perpendicular to the plates or parallel to the plates. This is accomplished by additives to the liquid crystal and/or surface treatments of the glass plates. The two orientations are termed homeotropic (perpendicular) and homogeneous (parallel) as indicated in Fig. 11.12. Depending on the materials and details of the configuration used, these cells will create displays based on various effects.

Nematic liquid crystals are diamagnetically anisotropic. By applying a magnetic field of ~2000 G, the longitudinal magnetic axes of the nematic crystals are aligned parallel to the magnetic field. Now, if an electric field is applied once parallel and once perpendicular to the longitudinal axes, two different dielectric constants are measured such that the difference is given by

$$\Delta \varepsilon = \varepsilon_{\parallel} - \varepsilon_{\perp}$$

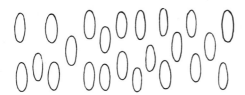

HOMEOTROPIC ORIENTATION

HOMOGENEOUS ORIENTATION

Figure 11.12 Orientations of nematic crystals (from Macklin, 1980). Reprinted from *The Optical Industry and Systems Purchasing Directory*, 1980.

where $\Delta\epsilon$ can be positive or negative depending on the type of molecules.

For molecules with negative $\Delta\epsilon$, the deformation of aligned phases (DAP) effect can be observed in liquid crystal cells with homeotropically oriented layers. When an increasing voltage is applied to the electrodes, molecular alignment approaches the parallel condition and may be observed by placing the liquid crystal cell between crossed polarizers.

If no voltage is applied, the display is opaque to light. As the applied voltage across the electrodes is increased, the entire spectrum of visible colors appears due to interference effects of the ordinary rays and the extraordinary rays of the birefringent crystals. When the voltage is removed, the molecules return to their original homeotropic orientation.

One can prepare a twisted nematic display by surface treatments of the opposing glass plates where the longitudinal axes of the molecules at one face are displaced 90° from those at the opposite face as indicated in Fig. 11.13. Such a twisted layer will rotate linearly polarized light 90°. When the arrangement is placed between crossed polarizers it appears transparent. If the layer consists of dielectrically positive liquid crystals, then an applied voltage will alter the molecular arrangement from the homogeneous to the homeotropic state. Now polarized light is not affected and the system is opaque. Such twisted nematic crystals are widely used in digital watches and many other digital instruments.

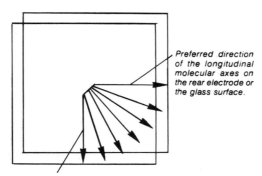

Preferred direction of the longitudinal molecular axes on the rear electrode or the glass surface.

Preferred direction of the longitudinal molecular axes on the front electrode or glass surface.

Figure 11.13  Twisted nematic layer (from Macklin, 1980). Reprinted from *The Optical Industry and Systems Purchasing Directory*, 1980.

## 11.6 THE LIQUID CRYSTAL LIGHT VALUE (LCLV)

Bleha and Robusto (1981) have described the LCLV. It is an optical-to-optical image transducer that is capable of accepting a low intensity, white or green input light image and converting it, in real time, into an output image by means of light from a second source. The input and output light beams are completely separate, and noninteracting. The device is shown schematically in Fig. 11.14.

The LCLV consists of a CdS photoconductor film and a nematic liquid crystal layer separated by a light blocking layer and a dielectric, borad band mirror. A biphenyl liquid crystal layer, 2 to 6 μm thick, is sandwiched between a transparent electrode and the dielectric mirror. The mirror reflects the readout light while the light blocking layer prevents residual readout light from reaching the photosensor. An ac voltage bias is imposed across the device. When there is no imaging light, most of the bias voltage falls

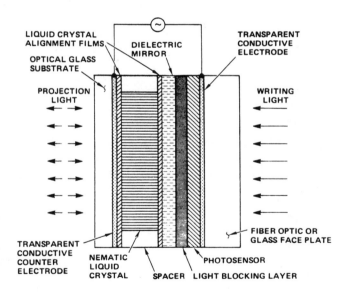

Figure 11.14 Cross-section of the LCLV module (from Bleha and Robusto, 1981). W. P. Bleha and P. F. Robusto, Optical-to-optical image comversion with the liquid crystal light valve, *Integrated Optics and Millimeter and Microwave Integrated Circuts*, Proc. SPIE *317*, 179-184, 1981.

across the ZnS while the voltage drop across the liquid crystal is below the activation threshold. When a light signal reaches the ZnS, its impedance drops and the voltage across the liquid crystal increases, thus activating the liquid crystal in a pattern that replicates the input image intensity.

In this reflection mode light valve arrangement, the nematic crystals are twisted through a 45° angle rather than the conventional 90°. The valve is dark during the off state. This arrangement requires that the readout light be polarized. An analyzer on the output face of the liquid crystal creates the intensity modulation in the output imagry. A beam splitter is used to direct the readout light and the liquid crystal light valve output.

## REFERENCES

A. Ballato, T. Lukaszek, Waves in piezoelectric crystals for frequency control and signal processing. Proc. SPIE *239*, Bellingham, WA, 1980.

J. M. Bennett and H. E. Bennett, Polarization, *Handbook of Optics*, W. G. Driscoll and W. Vaughan, Eds., McGraw-Hill Book Company, Inc., New York, 1978.

W. P. Bleha and P. F. Robusto, Optical-to-optical image conversion with the liquid crystal light valve, SPIE *317, Integrated Optics and Millimeter and Microwave Integrated Circuts*, SPIE, Bellingham, WA, 1981.

R. Goldstein, Pockels cell primer, *Laser Focus*, PennWell Publishing Co., Littleton, MA, 1968.

E. Hartfield and B. J. Thompson, Optical modulators, *Handbook of Optics*, W. G. Driscoll and W. Vaughan, Eds., McGraw-Hill Book Company, Inc., New York, 1978.

Interactive Radiation, Inc., *100 Series Miniature Electro-optic Modulators*, Northvale, NJ, 1980.

R. L. Macklin, Make-up of liquid crystals, *The Optical Industry and Systems Purchasing Directory*, Optical Publishing Co., Inc., Pittsfield, MA, 1980.

J. R. Meyer-Arendt, *Introduction to Classical and Modern Optics*, Prentice-Hall, Inc., Englewood Cliffs, NJ, 1972.

D. A. Pinnow and R. W. Dixon, Alpha-iodic acid:  a solution-grown crystal with a high figure of merit for acousto-optic device applications, Appl. Phys. Lett., *13*, p156, 1968.

A. Yariv, *Introduction to Optical Electronics*, Holt, Rinehart, and Winston, Inc., New York, 1971.

# GLOSSARY

**Abbe number**   Ratio of $(n_d - 1)/n_F - n_C)$, where the n's are the indices of refraction at the spectral lines corresponding to the subscripts; a measure of the medium's ability to bend the light relative to the dispersion.

**Absorption coefficient**   A measure of the attenuation of the intensity of a light ray as it passes through a medium as given by Bouguer's law.

**Absorption filter**   A medium which changes the spectral nature of light by reducing the energy content of the beam as it passes through the medium.

**Absorption index**   A material constant which is proportional to the product of the wavelength and the absorption coefficient.

**Achromatic lens**   A lens corrected for chromatic aberration for at least two selected colors or wavelengths.  It consists of at least two components.

**Acousto-optics**   The branch of optical science which deals with the effect of acoustic waves in an optical medium on the optical properties of that medium.

**Amphoteric**   Relating to a semiconductor, which can exhibit either p- or n-type properties.

**Analyzer**   An optical element that transmits only plane-polarized light.  An example is a properly cut and oriented calcite crystal.

**Anion**   An ion having a negative charge, i.e., an excess of electrons.

**Annealing**   A thermal process where a material is heated to a specified temperature, held there for a length of time sufficient to relieve all stresses, and then slowly cooled.

Antireflection coating   A film of material applied to an optical element which is designed to minimize reflection from the surface and enhance the light energy transmitted through the element.

Band gap   Electrons in crystals, according to the quantum theory, may exist at discrete energy levels only.  These levels are close together and form bands.  In a metal, these bands overlap and form a continuum.  In an electrical insulator, the bands are restricted and form an upper band (the conduction band) and a lower band (the valence band) in which all the allowed energy states are filled, with an intermediate range of energies in which no electrons may exist.  The difference in energy between the lowest energy level in the upper region and the highest permissible level in the lower band is termed the energy gap or band gap, usually expressed in electron volts.  In a semiconductor the band gap is small and electrons from the valence band are activated thermally into the conduction band.

Bandpass   That portion of the optical spectrum which is transmitted through a given optical element or system.

Band structure   The detailed description of the positions and energy levels of the energy bands described under Band gap.

Batching   In the making of glass, batching is the term applied to the weighing, mixing, preparing, and delivering to the glass melter the raw materials which will constitute the melt.

Beam splitter coating   A coating system applied to a substrate which optically separates the incoming optical beam into two or more frequency bands, one band being reflected and the second transmitted through the optical element.

Birefringent   Applied to a material which has different values of the index of refraction in different crystallographic directions.

Bouguer's law   The function which describes the attenuation of a light beam as it is transmitted through an optical material showing an exponential relationship of the intensity with respect to thickness.

Bravais lattice   A geometrical construction which describes the symmetry of a crystal.  The fourteen Bravais lattices are the only symmetries possible.  These Bravais lattices are the only shapes which, individually, can fill all space by replication.

Bridgman method   A method for growing single crystals which employs a furnace having a sharp thermal gradient through which a boat containing a molten charge is slowly moved, with the charge solidifying as it passes from the hot region to the cold region of the thermal gradient. The boat cavity has a sharp point where the first material to solidify forms a single crystal and the rest of the charge solidifies epitaxially to form the macroscopic single crystal. Alternatively, a single seed crystal can be employed in the leading edge of the boat upon which epitaxial growth takes place.

Canada balsam    An adhesive used to cement optical elements together. It is made from the sap of the North American balsam fir tree.

Cathode sputtering    A method of activating the source in a coating process where argon ions are electrostatically accelerated and bombard the source or target material.

Cation    An ion deficient in electrons which thus gives the ion a net positive charge.

Ceramic    A rather generalized term which refers to a class of materials which are compounds of metals and nonmetals, have relatively high melting or dissociation temperatures, and tend to be relatively hard.

CER-VIT®    A commercial glass-ceramic which has near zero coefficient of expansion near room temperature, manufactured by Owens Illinois Corporation.

Christiansen filter    An optical filter which is based on the suspension of a solid in a liquid in which the dispersion of the two materials is widely different but in which the index of refraction coincides for a given bandpass at one temperature. Scattering limits the specular transmission through the filter. The bandpass changes as the temperature varies.

Chromatic aberration    The departure from the ability of a lens to focus light of all colors sharply. The greater the dispersion of the lens material, the greater the chromatic aberration.

Coefficient of expansion    The proportionality factor between temperature change and dimensional change of a material due to that temperature change.

Complex refractive index    The refractive index has a real part and an imaginary part, based on the solution of Maxwell's equations which describe light passing through a material. The complex refractive index includes both the real and the imaginary terms.

The ordinary value of the refractive index usually stated is the real part of the complex refractive index.

Conduction band  As described under Band gap, electrons in a solid reside in certain allowed bands. The conduction band is the higher energy band within which electrons can migrate under the influence of an electric potential, giving rise to electrical current.

Cordierite  A lithium alumino-silicate ceramic crystalline compound which has a specific formulation and crystal structure. It has a relatively low coefficient of thermal expansion and is an important constituent of some low-expansion glass ceramics.

Covalent crystal  A crystal in which the interatomic bonding consists of an electron pair shared by neighboring atoms. Covalent crystals are relatively strong and hard. Diamond is an example of a crystal with a covalent bond.

Crown (glass)  A class of optical glass characterized by a high Abbe number and a low index of refraction. These glasses consist mainly of $SiO_2$, alkali metal oxides, and alkaline earth oxides.

Crown (furnace)  The roof of a glass melting furnace which is constructed in the form of an arch to be self-supporting. The arch becomes hot and reradiates energy to the melt.

Crystal  A material characterized by repeating patterns of atoms in three dimensions. The repeating pattern or symmetry of the crystal is controlled by interatomic forces.

Crystallographic axis  A direction in a crystal around which a symmetry of the atomic pattern exists.

Cut-on  In the transmission spectrum of an optically transmitting substance, the shortest wavelength at which detectable transmission occurs.

Cut-off  In the transmission spectrum of an optically transmitting substance, the longest wavelength at which detectable transmission occurs.

CVD process  Acronym for the chemical vapor deposition process in which solid substances are made by the reaction of gases in a processing chamber with a controlled atmosphere. The gases react at the surface of a substrate, and the deposit takes the form of the substrate. The substrate can be removed subsequently to leave a free-standing piece.

Czochralski process    A method of making single crystals in which a seed crystal is held on a rotating mandrel and brought in contact with the melt of the compound. As the seed is slowly withdrawn, a single crystal is formed by epitaxial growth.

Decibel [dB]    A measure of the intensity attenuation of a light beam as it passes through a medium such as an optic fiber. It is expressed as ten times the common logarithm of the ratio of the initial intensity to the intensity after traversing a unit length of fiber; for example, dB/km.

Density    A property of a material defined by the mass per unit volume of the substance.

Devitrification    The process whereby an amorphous substance (such as glass) converts to a crystalline form of the compound. In making optical glass the avoidance of devitrification is of paramount importance.

Diamagnetic    Referring to a material in which the electrical charges of the electronic structure tend to shield the interior of the body from an applied magnetic field. This gives the material an apparent magnetic permeability less than that which exists in a vacuum.

Diamond    A crystal composed entirely of covalently bound carbon atoms in a cubic lattice. It is the hardest substance known. Diamonds are said to be a girl's best friend.

Diamond point machining    A method of removing material from a body which uses diamond as the cutting bit and in which very accurately controlled and extremely fine cuts can be taken. Very precise machine tools are required, usually operating with gas or liquid bearings, and finishes as good as 10 Å rms can be achieved.

Dichroic coating    An optical coating which separates a light beam into distinct bands such as heat/light or color, reflecting one band while transmitting the other.

Dielectric constant    A material constant measured by comparison of the charge-holding ability of a capacitor with the material between the plates and that of the same capacitor with vacuum between the plates. The dielectric constant is given by the ratio of the two values.

Dielectric material    A material with a high band gap.

Dielectric stack    A multilayered optical coating which employs various types of dielectric materials for the various layers.

Diffraction    In optics, the bending of light as the light passes a
   boundary of an optical element. Diffraction effects can be ex-
   plained in terms of the destructive and constructive interference
   of light waves.

Dispersion    The effect on a beam of light as it passes through an
   optical medium due to the variation of the index of refraction as
   a function of wavelength. A simple example is the separation
   (or dispersion) of a beam of white light into colors by passing
   it through a glass prism.

E-beam    A beam of electrons used in a coating device where the
   E-beam impinges on and heats a target so that the target material
   evaporates and subsequently condenses on the substrate to be
   coated.

Electro-optics    That branch of optical science which deals with
   the effects of applied electrical voltage on the optical properties
   of optical materials.

Electromagnetic spectrum    The entire range of radiation, which is
   a consequence of oscillating electric charges, from long-wave
   radio waves to short-wave gamma rays and including the optical
   region of infrared, visible and ultraviolet. The electromagnetic
   spectrum is represented in terms of frequency, wavelength, or
   quantum energy.

Electronic structure    The configuration of electrons around the
   nucleus of an atom.

Extraordinary ray    When a beam of unpolarized light is incident
   on a birefringent crystal, such as calcite, there are two refracted
   beams instead of one as is found in a nonbirefringent material.
   One is called the ordinary (O) ray and the other the extraordinary (E) ray. The O ray behaves just as one would expect in an
   isotropic material, while the E ray behaves differently, changing
   direction even for a zero angle of incidence. The O and E rays
   are linearly polarized in planes orthogonal to each other.

Extrinsic absorption    That portion of the optical absorption in an
   optical material that is not inherent to a perfect single crystal
   or completely homogeneous amorphous specimen of the material.
   Extrinsic absorption may be due to impurities, voids, second
   phases, or grain boundary effects in polycrystalline substances.

Filter    An optical device which removes unwanted portions of the
   spectrum of an incident beam of light or separates the beam into
   two or more bands.

Fining   That part of the glassmaking process in which the molten glass is held at an appropriate temperature for a given length of time in a controlled atmosphere to improve the homogeneity of the glass just prior to forming.

Flint   That class of optical glass characterized by a relatively high refractive index and a low Abbe number. Flint glasses are generally composed of silica, lead oxide, and alkali metal oxides.

Float polishing   The process of polishing optical surfaces which employs very fine polishing particles in a fluid and depends on the collisions of these particles with the topmost surface of the sample without introducing subsurface damage. Using this process, 3 Å rms finishes have been achieved.

Floating zone method   A technique for growing single crystals whereby a narrow zone of metal is melted by the use of a ring induction heater around a solid bar. The melted zone is held in place by surface tension and is slowly moved along the length of the bar. A single crystal of material is epitaxially grown from the initially solidified material.

Fluorescence   The process whereby energy is absorbed by a body and then reemitted as visible or near visible light within $10^{-8}$ sec of the excitation.

Flux   Materials added to a batch (in glassmaking) which reduces the melting temperature, thus promoting fusion. An example is the addition of sodium oxide (soda ash) to silica.

Forming   The step in making a glass object where the melted glass is poured into a mold or some other device to create a desired shape.

Fracture surface energy   The amount of energy required to produce a given surface area of separation in a body under stress.

Fracture toughness   That property of a solid which is a measure of the resistance to crack propogation in that body.

Fresnel reflection   The reflection from the interface of two optical materials having different indices of refraction. The amount of reflection of the incident light is dependent on that difference as well as the angle of incidence.

Fresnel relations   Laws of reflection derived by Augustin Fresnel (1788−1827). Fresnel derived these laws, which specify the relations among the amplitudes of the electric vectors in the incident, reflected, and refracted light for light incident on the interface between two media with different indices of refraction.

Fused quartz    A glass made by melting crystalline quartz sand and then cooling the glass rapidly enough so that no devitrification takes place. An amorphous glass is the resultant product.

Fused silica    An amorphous silica glass made from high-purity precursors by the chemical vapor deposition (CVD) process.

Glass    An inorganic product of fusion which has cooled to a rigid body without crystallizing.

Glass ceramic    That class of materials which is composed of a glassy matirx within which microscopic crystals have precipitated. The crystalline component enhances the strength and fracture toughness of the material. Glass ceramics are made by a fusion process and cooling to an amorphous solid. Subsequent heat treatment is used to develop the crystalline phase. Generally, nucleating agents are used to control the degree of crystallinity.

Glow discharge cleaning    A method of cleaning the surface of an optical element prior to coating which employs a plasma gas. The surface is bombarded with energetic electrons, positive ions, chemically active atoms and molecules, neutral molecules, and high thermal energy and radiation.

Haze    A scattering effect in a lens caused by surface or other optical defects not large enough to be seen by the unaided eye.

Homeotropic orientation    The condition in a liquid crystal cell where the elongated molecules of the crystal are perpendicular to the glass plates comprising the cell.

Homogeneous orientation    The condition in a liquid crystal cell where the elongated molecules are parallel to the glass plates comprising the cell.

Intensity    For a light beam, the light energy power per unit area where the area is transverse to the propagation direction.

Interference filter    An optical filter composed of multilayers so selected and applied to the substrate that predetermined bands of the optical spectrum are removed by destructive interference.

Ionic    Relating to the properties of ions. Ions are either anions or cations.

Irtran®    A trade name of the Eastman Kodak Co., applied to a family of commercially available infrared transmitting materials.

Isostatic press    A device used in the fabrication of ceramics as well as other materials which applies hydrostatic or gas pressure to a batch of material enclosed in an impermeable sheath in order to densify or consolidate it. The isostatic pressure can be applied cold (cold isostatic press—CIP) or hot (hot isostatic press— HIP).

Junction laser.  A laser made up of an n-type semiconductor and a $p$-type semiconductor joined together to form a junction at the $n—p$ interface. By driving a large current through the junction transversely to the plane of the junction, the interface is made to lase.

Knoop hardness    A parameter which describes the hardness of a substance. The Knoop hardness is determined by indenting the material with a diamond indentor of a specified geometry and under a specified load. The size of the indent is used in a standardized formula to calculate the Knoop hardness number.

Laser    The acronym for light amplification by stimulated emission of radiation. The name laser is given to any device which can amplify light by the laser process.

Laser rod    A component of a laser made of a solid host such as a crystal or glass doped with an ion which can cause laser action to proceed under certain conditions. A common laser rod is ruby, which is a single crystal sapphire doped with approximately 0.3% of $Cr_2O_3$.

Lattice    A geometrical construct which describes the symmetries of crystals. See Bravais lattice.

Lattice parameter    In a lattice, the dimensions and angles between the edges which define the relative positions of the atoms that constitute the crystal characterized by the lattice.

Liquid crystal    A liquid in which long-range order of the atoms persists. These liquid crystals can be composed of linear, elongated molecules (strands) or two-dimensional arrays (planar). Certain liquid crystals possess the property that the boundaries of the elongated molecules do not interact with each other, giving rise to various magnetic and optical effects which have been found useful in display technology.

Longitudinal modulator    An electro-optic device in which a voltage is applied to the active crystal parallel to the direction of light. As the voltage is varied the optical properties of the crystal are varied or modulated.

Loss tangent    A property of a dielectric material associated with its dielectric constant and indicating the internal energy losses which occur as an alternating voltage is applied to the material.

Luminescence   The process whereby energy is absorbed in matter and subsequently reemitted as visible or near visible radiation. If the reemission occurs within $10^{-8}$ sec of the absorption, the process is called fluorescence; if longer, phosphorescence.

Maxwell's theory   The cornerstone of modern optical and electronic physics expressed by four partial differential equations which link all electromagnetic phenomena.

Melting   The process whereby energy is transferred to a material sufficient to cause molecular and/or atomic motion great enough to break the electronic bonding forces among them and to cause the material to assume a liquid state. This process is also called fusion.

Microyield strength   In a mirror substrate the stress required to produce a permanent degree of strain of a very low predetermined value. High microyield strength is desired to assure that the mirror figure experiences a minimum of permanent distortion due to normal stresses applied during its fabrication and application.

Mie scattering   Scattering caused by particles or other inhomogeneities in an optical medium which have dimensions on the order of the light wavelength of interest and of a different index of refraction than the medium.

Mirror substrate   The structural base upon which the figure and finishes of a mirror are applied.

Modulus of elasticity   In a body to which a stress is being applied, the ratio of the change in stress to the change in elastic strain. If the stress—strain relation is linear, the parameter is called Young's modulus. If the relation is nonlinear, then an instantaneous value of the modulus of elasticity can be computed for a specific value of applied stress.

Mullite   A crystal composed of a single-phase material with the average composition $3Al_2O_3 \cdot 2SiO_2$. The term is also applied to a furnace insulation block largely composed of the mullite crystal with a glassy binder.

Natural frequency   The frequency of vibration of an oscillating system in the absence of externally applied excitations.

Nematic phase   A state of matter in which a liquid crystal has the characteristic that the ends of the elongated molecules have little or no interactive forces with the ends of other molecules. Electrical and magnetic fields can alter the optical properties of nematic crystals.

Network former    In the formulation of glass, those additives which form agglomerations of atoms with short-range order so that the glass is not entirely amorphous. Networks lead to glasses with higher viscosities and aid in the forming process.

Neutral density filter    A filter which reduces the intensity of light passing through the filter equally over the entire bandpass.

Nucleation    The process during solidification from the melt whereby the initial crystal structures arise.

Optical glass fiber    A glass fiber which is prepared so that optical signals can be transmitted along the length of the fiber with minimal attenuation.

Ordinary ray    See Extraordinary ray.

Partial dispersion    The difference of the refractive index for a pair of wavelengths such as $n_C - n_b$, $n_d - n_C$, or $n_e - n_d$, where the subscripts refer to specific spectral lines.

Phase    A homogeneous portion of matter that is physically distinct and mechanically separable.

Phase equilibrium system    A group of phases related to each other by composition and equilibrium temperature. A phase equilibrium diagram for a two-component system (such as $SiO_2$-$Na_2O$) plots equilibrium temperature as the ordinate and composition as the abscissa. The diagram then divides itself into regions which display the composition and temperatures at which the various phases can exist.

Phonon    In solid state physics, the energy in a lattice vibration or elastic wave which has been quantized. The phonon is an analogy to the photon, which is the quantum of energy in an electromagnetic wave.

Piezo-optic constant    The piezo-optic effect is analogous to the electro-optic effect in that pressure rather than voltage applied results in a change in the birefringence of the crystal. The piezo-optic coefficient is a measure of the change in index of refraction for a given change in applied pressure. The piezo-optic effect is characterized by a tensor analogous to the electro-optic effect.

Piezoelectric    Refers to the effect in a solid whereby application of pressure induces a voltage. Quartz is an example of a piezo-electric crystal.

**Pockels effect**   Occurs in certain crystals where an applied voltage changes the degree of birefringence of the crystal. In the Pockels effect, the change in birefringence is linear with respect to the applied voltage. The Kerr effect is similar, except in that case the change in birefringence is quadratic with respect to the applied voltage.

**Poisson's ratio**   In mechanics, when a bar is subjected to a tensile stress, the inverse ratio of the elongation of the bar to the reduction in its cross-sectional linear dimension.

**Polarized light**   A light beam in which all of the electromagnetic waves are aligned with parallel magnetic vectors and parallel electric vectors. This type of light is described as plane-polarized.

**Polarizer**   An optical element which transforms a beam of unpolarized light into a beam of plane-polarized light.

**Polycrystalline**   Describes a body composed of many small crystals in which the individual crystals are called grains and are bonded to each other by a more or less amorphous grain boundary material.

**Principal dispersion**   The difference $n_F - n_C$ is called the principal dispersion. This value is used in the calculation of the Abbe number.

**Pyrex®**   A two-phase glass, consisting of two liquid phases in the melt persisting in the solid as amorphous phases, which has a low thermal coefficient of expansion and is strong. The product was developed by Corning Glass Works.

**Quantum theory**   A theory in physics which deals with the interaction of energy and matter in which electrons are permitted to have energy only at discrete values or quanta. The theory accepts the dual wave-particle character of electromagnetic energy.

**Rayleigh scattering**   Scattering of light by particles that have linear dimensions much smaller than the wavelength of light being scattered.

**Reflection**   When a beam of light is incident upon a surface, some of the light is returned in the general direction of the incoming beam without change of wavelength. The returned light is called the reflected beam, which may be specular or diffuse. This process is called reflection.

**Refractive index**   The ratio of the speed of light in vacuum to the speed of light in a given material.

Refractory    The description of a material with a set of properties which permit the material to withstand high temperature.

Reststrahlen reflection    A reflection peak from the surface of a dielectric crystalline material which occurs at the reststrahlen wavelength. The reststrahlen wavelength occurs at a frequency which excites the lattice in the crystal and causes the index of refraction to undergo a rapid variation.

Roughness [rms]    A measure of the smoothness of a surface. The root mean square (rms) is defined as the square root of the mean of the squares of the peak-to-valley variations in the surface.

Ruby    A crystal composed of aluminum oxide doped with chromium oxide. It has a characteristically red color.

Scattering    The process whereby light beams are deflected by reflection and diffraction more or less randomly from the incident direction. Scattering is caused by differences in the index of refraction between the main medium and its minor constituents such as dust or moisture in air or bubbles in glass.

Scratch-dig    The term applied to optical element surface defects such as scratches or pits (digs). Specifications have been developed which define the severity of scratches or digs allowable for a given class of optical elements.

Secondary spectrum    In an achromatic lens composed of two different glasses, two predetermined colors focus at the same place but all other colors are slightly out of focus. This small residual color defect is known as the secondary spectrum.

Selective scattering    A technique used in certain filters which function by removing unwanted parts of the spectrum by the scattering process. A typical example is the use of gold precipitates in a glass matrix which scatter all wavelengths below a given threshold, dependent on the second phase particle size.

Semiconductor    A crystalline substance having a relatively low band gap, in which electrons are thermally activated into the conduction band, thereby becoming capable of conducting electric charge. Semiconductors are less conductive than metals but more conductive than dielectrics.

Single crystal    A body composed of atoms symmetrically oriented in repeating patterns with no change in direction of the axes of symmetry and no interruptions by grain boundaries.

Sinter   A thermal process in which closely packed powders are consolidated by atomic and molecular diffusional events to form a dense, relatively homogeneous body.

Slip   A suspension of ceramic powders in water which is poured into a water-absorbing plaster mold. The shape formed is subsequently dried, removed from the mold, and sintered to form a ceramic body.

Specific heat   The amount of energy which must be absorbed by a material to raise its temperature by one unit, expressed in units of energy per unit of temperature per gram mass of material.

Spectral filter   A filter which will pass light in a narrow range of wavelengths.

Spectrophotometer   An optical device which can measure reflection and transmission through an optical element as functions of wavelength.

Specular   Refers to a beam of light which is not scattered.

Spinel   A crystalline form characterized by a cubic lattice and with a chemical formulation of $AB_2O_4$, where A is divalent and B is trivalent. An example is $Mg_2AlO_4$.

Staining   The process in which the surface of an optical element becomes discolored due to some form of environmental attack such as might be due to humidity or pollutants in the air.

Statistical fracture analysis   The evaluation of fracture data obtained from many samples of material made by the same process. Such a study leads to a prediction of the probability of failure of an element when subjected to a given set of stresses. This technique is of special importance in the design of ceramic structures because of the inherent brittleness of ceramic materials.

Stress optical coefficient   In an optical material subject to stress, it is found that the induced birefringence due to the stress is proportional to the stress. The stress optical coefficient is the proportionality constant.

Stress/strain birefringence   The birefringence in an optical material caused by the stress and consequent strain in the material.

Striae   Regions or veins of inhomogeneity in an optical glass where variations in index of refraction can be detected visually by the examination of the glass under polarized light.

Striking process   The process whereby precipitates are formed in a matrix for use as a filter employing selective scattering.

Superpolishing    See Float polishing

Tensile strength    That property of a material defined by its ability to resist stress without fracture. Tensile strength is given in stress per unit area at failure.

Tensor    A scalar quantity has only magnitude; a vector quantity has magnitude and direction. However, there are certain entities which cannot be represented by a scalar or a vector. The stretching of a rod by application of force at each end is an example of such an entity. In a three-dimensional continuum, let each rectangular component of a vector B be a linear function of the components of a vector A:

$$B_1 = T_{11} A_1 + T_{12} A_2 + T_{13} A_3$$

$$B_2 = T_{21} A_1 + T_{22} A_2 + T_{23} A_3$$

$$B_3 = T_{31} A_1 + T_{32} A_2 + T_{33} A_3$$

A tensor of rank two (described by the above set of equations) is defined as a linear transformation of the components of vector A into the components of vector B which is invariant to rotations of the coordinate system. The nine components of the linear transformation are called the tensor components or coefficients. An example of a second-rank tensor is electrical resistivity. A third-rank tensor has eighteen tensor components or coefficients. Examples of third-rank tensor properties are the piezoelectric effect and the electro-optic effect.

Thermal conductivity    That property of a material which defines the ability of thermal energy to pass through the material. It is expressed as units of energy passing per unit area transverse to the direction of energy flow per unit of thickness in the direction of flow per unit of time per unit of temperature difference across the unit thickness.

Thermal diffusivity    A material property defined as the thermal conductivity divided by the product of the density and the specific heat.

Thermal optic coefficient    The proportionality constant which defines the linear relation between a change in temperature of an optical medium and the change in its index of refraction.

Thermoplastic    A polymeric material which attains a rigid shape when placed in a mold and then cooled and, in addition, can be reheated after cooling to be reformed into another shape.

Thermoset   A polymeric material which attains a rigid shape when placed in a mold and then is heated to effect polymerization. Once shaped in this manner, a thermoset cannot be reshaped by the simple application of heat.

Torsional rigidity   The modulus of elasticity of an elastic material under the influence of shear.

Transformation temperature   The temperature at which a material changes phase. Common examples are the freezing temperature and the boiling temperature of water. The transformation also can be a change in crystal form such as occurs during the devitrification of glass.

Transition   In the band theory of solids, the transfer of an electron from one allowed quantum state to another.

Transmittance (external)   The ratio of the incident light energy power per unit area (intensity) to that of the exiting light energy power per unit area, including the reflection effects due to the boundaries of the medium. The areas are assumed transverse to the direction of propagation of the light beam.

Transmittance (internal)   For a beam of light in the medium, the ratio of the light intensity just after entering the medium to the intensity just before leaving the medium.

Transverse modulator   An electro-optic crystal so configured that its optical properties are varied by means of a voltage applied transversely to the direction of propagation of the beam.

ULE®   The trade name for an ultra-low expansion glass ceramic manufactured by Corning Glass Works. It has near zero coefficient of expansion, near room temperature, and is used for optical element substrates where extreme dimensional stability is essential, such as in telescopes.

V-coat   A multilayer optical coating designed to pass a narrow band of wavelengths.

Valence band   In the band theory of solids, the lower band of allowed energy levels which are filled by electrons.

Verneuil method   A technique for making a single crystal in which a gaseous vapor from a flame condenses epitaxially on a rotating seed crystal to form a larger crystal.

Vickers hardness   Similar to Knoop hardness, except that the hardness number is determined by the indentation of a diamond indentor in the form of a square pyramid as opposed to the Knoop indentor, which is diamond-shaped with an aspect ratio of 7:1.

Viscosity    The property of a fluid measured by its resistance to deformation under the influence of a shear force.

Volume resistivity    The electrical resistance of a unit cube of material where the voltage is applied across two parallel faces of the cube.

Weibull modulus    A parameter which defines the degree of variation of strength measurements in a large number of specimens, all made by the same process. The Weibull modulus is used in the probabilistic prediction of failure of materials and is most often applied to brittle materials which are prone to brittle failure.

Witness plate    A flat plate of the same composition and surface preparation as the substrates being coated in a batch-coating process. The film deposited on the witness plate is evaluated as being representative of what was deposited on the substrates.

YAG    Acronym for yttrium alumina garnet, $Y_3Al_5O_{12}$, which is used as a laser rod when appropriately doped, as with neodymium.

ZERODUR®    An ultra-low expansion mirror substrate manufactured by Schott Glass Technologies, Inc., similar to ULE®.

# INDEX